Introduction to
Solid Modeling
Using SolidWorks®

McGraw-Hill continues to bring you the *BEST* (**B**est **E**ngineering **S**eries and **T**ools) approach to introductory engineering education:

Chapman, *Fortran 90/95 for Scientists and Engineers,* 2/e ISBN 0072922389
Donaldson, *The Engineering Student Survival Guide,* 3/e ISBN 0073019259
Eide/Jenison/Northup, *Introduction to Engineering Design and Problem Solving,*
    2/e ISBN 0072402210
Eisenberg, *A Beginner's Guide to Technical Communication,* ISBN 0070920451
Finklestein, *Pocket Book of Technical Writing for Engineers and Scientists,*
    2e ISBN 0072976837
Gottfried, *Spreadsheet Tools for Engineers using Excel,* ISBN 0072480688
Howard/Musto, *Introduction to Solid Modeling Using SolidWorks®,* ISBN 0072978775
Palm, *Introduction to MatLab 7 for Engineers,* ISBN 0072922427
Pritchard, *Mathcad: A Tool for Engineering Problem Solving,* ISBN 0070121893
Schinzinger/Martin, *Introduction to Engineering Ethics,* ISBN 0072339594
Smith, *Teamwork and Project Management,* 2/e ISBN 0072922303
Tan/D'Orazio, *C Programming for Engineering and Computer Science,*
    ISBN 0079136788

Additional Titles of Interest:

Andersen, *Just Enough Unix,* 4/e ISBN 0072463775
Eide/Jenison/Mashaw/Northup, *Engineering Fundamentals and Problem Solving,*
    4/e ISBN 0072430273
Holtzapple/Reece, *Foundations of Engineering,* 2/e ISBN 0072480823
Holtzapple/Reece, *Concepts in Engineering,* ISBN 0073011770
Martin/Schinzinger, *Ethics in Engineering,* 4/e ISBN 0072831154

Introduction to
Solid Modeling
Using SolidWorks®

William E. Howard
Milwaukee School of Engineering

Joseph C. Musto
Milwaukee School of Engineering

 Higher Education

Boston   Burr Ridge, IL   Dubuque, IA   Madison, WI   New York   San Francisco   St. Louis
Bangkok   Bogotá   Caracas   Kuala Lumpur   Lisbon   London   Madrid   Mexico City
Milan   Montreal   New Delhi   Santiago   Seoul   Singapore   Sydney   Taipei   Toronto

![Mc Graw Hill] **Higher Education**

INTRODUCTION TO SOLID MODELING USING SOLIDWORKS®

Published by McGraw-Hill, a business unit of The McGraw-Hill Companies, Inc., 1221 Avenue of the Americas, New York, NY 10020. Copyright © 2006 by The McGraw-Hill Companies, Inc. All rights reserved. No part of this publication may be reproduced or distributed in any form or by any means, or stored in a database or retrieval system, without the prior written consent of The McGraw-Hill Companies, Inc., including, but not limited to, in any network or other electronic storage or transmission, or broadcast for distance learning.

Some ancillaries, including electronic and print components, may not be available to customers outside the United States.

This book is printed on recycled, acid-free paper containing 10% postconsumer waste.

1 2 3 4 5 6 7 8 9 0 QPD/QPD 0 9 8 7 6 5 4

ISBN 0-07-297877-5

Sponsoring Editor: *Suzanne Jeans*
Managing Developmental Editor: *Emily J. Lupash*
Marketing Manager: *Dawn R. Bercier*
Senior Project Manager: *Kay J. Brimeyer*
Senior Production Supervisor: *Sherry L. Kane*
Media Technology Producer: *Eric A. Weber*
Senior Designer: *David W. Hash*
Cover Designer: *Rokusek Design*
(USE) Cover Image: *Cover photo and SolidWorks® rendering provided by Mountain Cycle*
Compositor: *International Typesetting and Composition*
Typeface: *10/12 Times*
Printer: *Quebecor World Dubuque, IA*

Library of Congress Cataloging-in-Publication Data

Howard, William E.
    Introduction to solid modeling using SolidWorks® / William E. Howard, Joseph C. Musto.—1st ed.
        p. cm.
    Includes index.
    ISBN 0-07-297877-5
    1. Computer graphics. 2. SolidWorks®. 3. Engineering models. 4. Computer-aided design. I. Musto, Joseph C. II. Title.

T385.H75 2006
620'.0042'0285536—dc22

2004059201
CIP

www.mhhe.com

Contents

Motivation for This Text

As design engineers, the authors witnessed the development of solid modeling software from a distance. The cost and complexity of early solid modeling software limited its adoption, and companies using the software usually assigned specialists to learn and use it full-time. For most design engineers, solid modeling seemed to be an exotic tool that would not affect the way we did our jobs.

Later, as engineering professors, we both had the opportunity to learn and use SolidWorks® solid modeling software, and we were instantly hooked. Not only was the utility of the software obvious, but the program was easy to learn and fun to use. For the last six years, we have shared our enthusiasm with the program with hundreds of students, in classes and informal training sessions at Milwaukee School of Engineering, and in summer programs with high school students. Most of the material in this book began as tutorials that we developed for these purposes. We continue to be amazed at how quickly students at all levels can learn the basics of the program, and by the sophisticated projects that many students develop after only a short time using the software.

While anyone desiring to learn the SolidWorks program can use this book, we have added specific elements for beginning engineering students. With these elements, we have attempted to introduce students to the design process and to relate solid modeling to subjects that most engineering students will study later. We hope that the combination of the tutorial style approach to teaching the functionality of the software together with the integration of the material into the overall study of engineering will motivate student interest not only in the SolidWorks software but in the profession of engineering.

Philosophy of This Text

The development of powerful and integrated solid modeling software has continued the evolution of computer-aided design packages from drafting/graphical communication tools to full-fledged engineering design and analysis tools. A solid model is more than simply a drawing of an engineering component; it is a true virtual representation of the part, which can be manipulated, combined with other parts into complex assemblies, used directly for analysis, and used to drive the manufacturing equipment that will be used to produce the part.

This text was developed to exploit this emerging role of solid modeling as an integral part of the engineering design process; while proficiency in the software will be achieved through the exercises provided in the text, the traditional "training" exercises will be augmented with information on the integration of solid modeling into the engineering design process. These topics include:

- The exploitation of the parametric features of a solid model, to not only provide an accurate graphical representation of a part but also to effectively capture an engineer's design intent,
- The use of solid models as an analysis tool, useful for determining properties of components as well as for virtual prototyping of mechanisms and systems,
- The integration of solid modeling with component manufacturing, including the generation of molds, sheet metal patterns, and rapid prototyping files from component models.

Through the introduction of these topics, students will be shown not only the powerful modeling features of the SolidWorks program, but also the role of the software as a full-fledged integrated engineering design tool.

The Use of This Text

This text contains chapter-long tutorials that introduce both basic concepts in solid modeling (such as part modeling, drawing creation, and assembly modeling) and more advanced applications of solid modeling in engineering analysis and design (such as mechanism modeling, mold creation, sheet metal bending, and rapid prototyping). Each tutorial is organized as "keystroke-level" instructions, designed to teach the use of the software.

While these tutorials offer a level of detail appropriate for new professional users, this text was developed to be used as part of an introductory engineering course, taught around the use of solid modeling as an integrated engineering design and analysis tool. Since the intended audience is undergraduate students new to the field of engineering, the text contains features that help to integrate the concepts learned in solid modeling into the overall study of engineering. These features include:

- *Design Intent Boxes:* These are intended to augment the "keystroke-level" tutorials to include the rationale behind the sequence of operations chosen to create a model.

- *Future Study Boxes:* These link the material contained in the chapters to topics that will be seen later in the academic and professional careers of new engineering students. They are intended to motivate interest in advanced study in engineering, and to place the material seen in the tutorials within the context of the profession.

While these features are intended to provide additional motivation and context for beginning engineering students, they are self-contained, and may be omitted by professionals who wish to use this text purely for the software tutorials.

The Organization of This Text

The organization of the chapters of the book reflects the authors' preferences in teaching the material, but allows for several different options. We have found that covering drawings early in the course is helpful in that we can have students turn in drawings rather than parts as homework assignments. The eDrawings feature, which is covered in Chapter 2, is especially useful in that eDrawings files are small (easy to e-mail), self-contained (not linked to the part file), and can be easily marked up with the editing tools contained in the eDrawings program.

The following flowchart illustrates the relations between chapters, and can be used to map alternative plans for coverage of the material. For example, if it is desired to cover assemblies as soon as possible (as might be desired in a course that includes a project) then the chapters can be covered in the order 1-3-6-7-2-8, with the remaining chapters covered in any order desired. An instructor who prefers to cover parts, assemblies, and drawings in that order may cover the chapters in the order 1-3-4-5-6-7-2-8, again with the remaining chapters covered in any order.

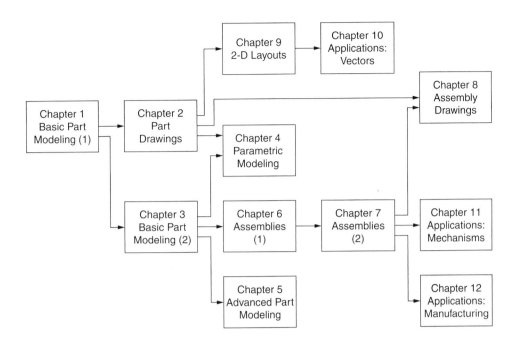

Chapters 9 and 10 may be omitted in a standard solid modeling course; however, they can be valuable in an introductory engineering course. Engineering students will almost certainly find use at some point for the 2-D layout and vector mechanics applications introduced in these chapters. Chapter 13 is intended to wrap up a course with a discussion of how solid modeling is used as a tool in the product development cycle.

Additional Resources for Instructors

Additional recourses are available on the web at www.mhhe.com/howardmusto. Included on the website are tutorials for two popular SolidWorks Add-Ins, COSMOSMotion™ and PhotoWorks™, and the book figures in PowerPoint format. Instructors can also access model files for all tutorials and problems (password-protected; contact your McGraw-Hill representative for access).

Acknowledgments

We are grateful to the many people at McGraw-Hill and SolidWorks Corporation for their help and encouragement. At McGraw-Hill, Tim Maruna persuaded us to pursue this project and has offered support at every step. Kelly Lowery, Emily Lupash, Dawn Bercier, Suzanne Jeans, and Amanda Green smoothly guided us through the processes of writing and editing. Carol Kromminga copyedited the manuscript and offered many helpful suggestions. Kay Brimeyer managed the production process, and helped us turn our manuscript into a finished product. At SolidWorks, Rosanne Kramer helped us to get started

"COSMOSMotion is a trademark of Structural Research and Analysis Corporation. PhotoWorks is a trademark of SolidWorks Corporation."

with the SolidWorks software and continues to help and support the use of the software at our institution. Magdalen Fryatt and the rest of the SolidWorks Partner Team were also very helpful.

Several reviewers offered comments that undoubtedly made the book better. These reviewers included:

Bruce H. Adee, *University of Washington*

Lawrence E. Carlson, *University of Colorado*

Karen L. Coen-Brown, *University of Nebraska, Lincoln*

Rajit Gadh, *University of California, Los Angeles*

Meung J. Kim, *Northern Illinois University*

Roger L. Ludin, *California Polytechnic State University, San Luis Obispo*

Sara McMains, *University of California, Berkeley*

Charles Ritz, *California Polytechnic State University, Pomona*

Mary Tolle, *South Dakota State University*

We thank these reviewers for their efforts.

At Milwaukee School of Engineering, many of our students and faculty colleagues have used early versions of the manuscript and materials that eventually became this text. We thank them for their patience and helpful feedback along the way.

Ed Howard

Joe Musto

December 2004

Basic Part Modeling Techniques

Introduction

Solid modeling has become an essential tool for most companies that design mechanical structures and machines. Just 15 years ago, this would have been hard to imagine. While 3-D modeling software existed, it was very expensive and required high-end computer workstations to run. An investment of $50,000 or more was required for every workstation with software, not including training of the operator. As a result, only a few industries used solid modeling, and the trained operators tended to work exclusively with the software.

The dramatic performance improvements and price drops of computer hardware, along with increased competition among software vendors, have significantly lowered the cost barrier for companies to enter the solid modeling age. The software has also become much easier to use, so that engineers who have many other job functions can use solid modeling when required without needing to become software specialists. The SolidWorks® program was among the first solid modeling programs to be written exclusively for the Microsoft Windows environment.[1] Since its initial release in 1995, it has been adopted by thousands of companies worldwide.

This text is laid out as a series of tutorials that cover most of the basic features of the SolidWorks® program. Although these tutorials will be of use to anyone desiring to learn the software, they are written primarily for freshman engineering students. Accordingly, topics in engineering design are introduced along the way. "Future Study" boxes give a preview of coursework that engineering students will encounter later, and relate that coursework to the solid modeling tutorials.

In this first chapter, we will learn how to make a simple part in SolidWorks. We will actually create the same part twice, using different techniques the second time.

In this chapter, you will:

- be introduced to the process of engineering design and the concept of concurrent engineering,
- learn how to create and dimension simple 2-D sketches and perform operations to create 3-D shapes from these sketches,
- be introduced to design intent as we create a part, and
- learn how to modify parts.

1.1 Engineering Design and Solid Modeling

The term *design* is used to describe many endeavors. A clothing designer creates new styles of apparel. An industrial designer creates the overall look and function of consumer

---

[1] SolidWorks is a registered trademark of the SolidWorks Corporation, 300 Baker Avenue, Concord, MA 01742.

products. Many design functions concentrate mainly on *aesthetic* considerations—how the product looks, and how it will be accepted in the marketplace. The term *engineering design* is applied to a process in which fundamentals of math and science are applied to the creation or modification of a product to meet a set of objectives.

Engineering design is only one part of the creation of a new product. Consider a company making consumer products, for example bicycles. A marketing department determines the likely customer acceptance of a new bike model and outlines the requirements for the new design. Industrial designers work on the preliminary design of the bike to produce a design that combines functionality and styling that customers will like. Manufacturing engineers must consider how the components of the product are made and assembled. A purchasing department will determine if some components will be more economical to buy than to make. Stress analysts will predict whether the bike will survive the forces and environment that it will experience in service. A model shop may need to build a physical prototype for marketing use or to test functionality.

During the years immediately following World War II, most American companies performed the tasks described above more or less sequentially. That is, the design engineer did not get involved in the process until the specifications were completed, the manufacturing engineers started once the design was finalized, and so on. From the 1970s through the 1990s, the concept of *concurrent engineering* became widespread. Concurrent engineering refers to the process in which engineering tasks are performed simultaneously rather than sequentially. The primary benefits of concurrent engineering are shorter product develop times and lower development costs. The challenges of implementing concurrent engineering are mostly in communications— engineering groups must be continuously informed of the actions of the other groups.

Solid modeling is an important tool in concurrent engineering in that the various engineering groups work from a common database: the solid model. In a 2-D CAD (Computer-Aided Design) environment, the design engineer produced sketches of the component, and a draftsman produced 2-D design drawings. These drawings were forwarded to the other engineering organizations, where much of the information was then duplicated. For example, a toolmaker created a tool design from scratch, using the drawings as the basis. A stress analyst created a finite element model, again starting from scratch. A model builder created a physical prototype by hand from the drawing parameters. With a solid model, the tool, finite element model, and rapid prototype model are all created directly from the solid model file. In addition to the time savings of avoiding the steps of re-creating the design for the various functions, many errors are avoided by having everyone working from a common database. Although 2-D drawings are usually still required, since they are the best way to document dimensions and tolerances, they are linked directly to the solid model and are easy to update as the solid model is changed.

A mechanical engineering system (assembly) may be composed of thousands of components (parts). The detailed design of each component is important to the operation of the system. In this chapter, we will step through the creation of a simple component. In future chapters, we will learn how to make 2-D drawings from a part file, and how to put components together in an assembly file.

1.2 Part Modeling Tutorial: Flange

This tutorial will lead you through the creation of a simple solid part. The part, a flange, is shown in Figure 1.1 and is described by the 2-D drawing in Figure 1.2.

Begin by clicking the SolidWorks icon. The "Welcome to SolidWorks" box shown in Figure 1.3 may be displayed. If so, click the "Show at startup" box (the check mark will

FIGURE 1.1

FIGURE 1.2

Scale = 1:2
All dimension are inches and degrees
Fillets are .25 radius

FIGURE 1.3

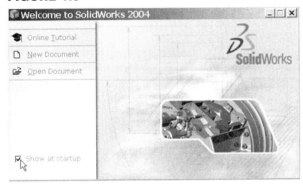

be removed) and close the box by clicking the X in the upper-right corner. From the main menu, select File: New, as shown in Figure 1.4.

FIGURE 1.4

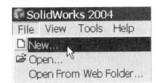

A dialog box will appear, from which you are to specify whether we are creating a part, an assembly, or a drawing.

Click on the Part icon to highlight it (see Figure 1.5), and then click the OK button.

FIGURE 1.5

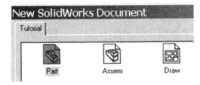

A box called the Quick Tips box, shown in Figure 1.6, may appear in the lower right of the screen. We will not be using the Quick Tips in this book, so you can turn off the display of this box.

Select Help from the main menu, and click on Quick Tips, as shown in Figure 1.7.

Before we begin modeling the flange, we will establish a consistent setup of the SolidWorks environment. If you right-click on any of the toolbars, a list of available toolbars is displayed, with active toolbars indicated by a check mark.

FIGURE 1.6

FIGURE 1.7

Right-click on the main menu or on any of the toolbars at the top of the screen. On the list of toolbars that is displayed, click to select or deselect toolbars so that the Standard, Standard Views, and View toolbars are selected, as shown in Figure 1.8.

The toolbars may be repositioned anywhere on the screen by clicking and dragging with the left mouse button. The layout shown in Figure 1.9 is similar to the default configuration. The layout of the part modeling screen also includes a main menu at the top of the screen, and a status bar on the bottom of the screen. When you move the cursor over any toolbar icon or menu command, a message on the left side of the status bar describes the command. Other information, such as whether or not a sketch is completely defined, appears at the right side of the status bar.

FIGURE 1.8

FIGURE 1.9

The CommandManager contains most of the tools that you will use to create parts. By default, there are two categories of tools in the CommandManager: sketch tools used in creating 2-D sketches, and features tools used to create and modify 3-D features. Clicking on the Sketch and Features icons at the left of the CommandManager toggles the commands between the 2-D and 3-D tools, as shown in Figures 1.10 and 1.11.

Sketch

Features

Just to the left of the drawing area is the FeatureManager Design Tree. The steps that you will execute to create the part will be listed in the FeatureManager. This information is important when the part is to be modified. When you open a new part, the FeatureManager lists an origin and three predefined planes. As you select each plane with your mouse, the plane is highlighted in the drawing area. We can create other planes as needed, and will do so later in this tutorial.

FIGURE 1.10

FIGURE 1.11

FIGURE 1.12

FIGURE 1.13

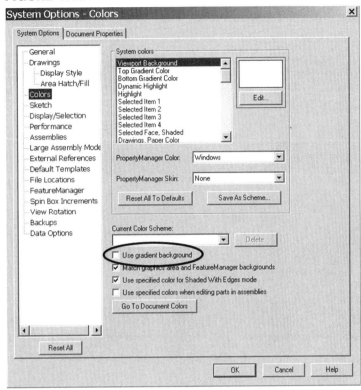

Before we get started, let's make sure that some of the program options are set the way we want them.

From the menu, select Tools: Options, as shown in Figure 1.12.

The dialog box contains settings for both the system and for the specific document that is open. By default, the background color of the modeling area is blue, with a color gradient to white. If you would prefer a plain white background, follow the next step.

Under System Options and Colors, click on Viewport Background and Edit. Pick white as the background color, and click OK. Clear the box labeled "Use gradient background," as shown in Figure 1.13. Under the Document Properties tab, select Detailing. Under Dimensioning standard, select ANSI from the pull-down menu, as shown in Figure 1.14.

FIGURE 1.14

FIGURE 1.15

FIGURE 1.16

Also under the Document Properties tab, select Grid/Snap and check or uncheck the boxes as shown in Figure 1.15.

With these settings, a grid will appear when we are sketching, but automatic snapping to points will not occur.

Under Units, select IPS (inch, pound, second) as the unit system. Set the number of decimal places for inches to 3, as shown in Figure 1.16.

Finally, under Colors, click on Shading, as shown in Figure 1.17, and Edit. Select a color for the model. A relatively light color is recommended. Click OK to close the

FIGURE 1.17

FIGURE 1.18

color dialog box, and OK again to close the Options box.

Any of the options just set can be changed at any time during the modeling process.

We start the construction of the flange by sketching a circle and extruding it into a 3-D disk.

FIGURE 1.19

Select the Front Plane by clicking on it in the FeatureManager Design Tree, as shown in Figure 1.18.

The Front Plane will be highlighted in green, as shown in Figure 1.19. The color green indicates that an item is the currently selected item.

Begin a sketch by selecting the Sketch group of the CommandManager, and then the Sketch Tool.

 Sketch

 Sketch

FIGURE 1.20

Since the Display grid option was selected earlier under grid/snap, a grid pattern appears, signifying that you are in the sketching mode. Also, a Sketch icon appears in a triangle at the upper-right corner of the screen, as shown in Figure 1.20. (Note: For clarity, most of the images in this text were produced with the grid display turned off. For new users, however, the grid display is a useful reminder of whether the sketching mode is active or not.)

Select the Circle Tool.

 Circle

Notice as you move the cursor into the drawing area that it changes appearance into a pencil with a circle next to it, as shown in Figure 1.21. This lets you know that the Circle Tool is active.

FIGURE 1.21

Move the cursor toward the origin until a box appears next to the pencil, as shown in Figure 1.22; this indicates that you will snap to an existing point (in this case the origin when you click with the mouse).

FIGURE 1.22

A *snap* is the implied relation of the positions of two entities. In this example, when you "snap" to the origin, the circle will be centered at the exact coordinates of $x = 0$ and $y = 0$. The relation added when one entity is created by snapping to another can be edited later, if desired. The addition of a snap automatically is a nice feature of the SolidWorks program: snaps are intuitive. It is not necessary to enter the numerical coordinates of the center of the circle.

FIGURE 1.23

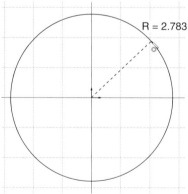

R = 2.783

Now click and hold the left mouse button and drag the mouse outward to create a circle, as shown in Figure 1.23.

Notice that the radius of the circle appears next to the cursor. Don't worry about the dimension now; just release the mouse to complete your circle. At this point, the Circle Tool is still active, as indicated by its green color.

Click the esc key to close the Circle Tool and deselect the circle just drawn.

The circle should now appear in blue. At the bottom of the screen, notice that "Under Defined" appears. This is because we have not set the dimension of the circle yet. When a sketch does not contain enough dimensions and/or relations to define its size and position in space, it is said to be underdefined, and is denoted by blue entities. Other possible conditions of the sketch are "Fully Defined," when the sketch contains exactly enough dimensions and/or relations to define its size and position in space (denoted by black entities), and "Over Defined," where the sketch has at least one dimension or relation that contradicts or is redundant to the other dimensions and relations (denoted by red entities). Overdefined sketches should be avoided.

FIGURE 1.24

Select the Smart Dimension Tool. Now click with the left mouse button anywhere on the circle. A dimension will be added to the diameter of the circle. You can drag the dimension to a convenient location. When the dimension is where you want to put it, left click the mouse (see Figure 1.24).

FIGURE 1.25

A dialog box prompting for this dimension will be displayed, as shown in Figure 1.25. Enter "5.5" in the box. (You don't need to enter "in" since that is the default unit, set earlier). Press the Enter key or click on the check mark to update the dimension. Notice that the circle is redrawn to the correct dimension.

The dimension in inches will be displayed, and the circle will become black, as shown in Figure 1.26. Notice at the bottom of the screen in the status bar that the sketch is now Fully Defined. (Note: If we had not snapped to the origin for the circle's center, the sketch would still be underdefined because the circle would not be located in space.)

FIGURE 1.26

Use the esc key to turn off the Dimension Tool. Now if you double-click on the dimension value, the dialog box reappears and you can change the dimension. Try this, and then use the Undo Tool to return the dimension to 5.5 inches.

Note that you can change the way the dimension is displayed. Select the dimension by clicking once on the dimension value, and then click on one of the green dots on the arrow heads to change the arrow display. Click and drag on the center green dot to rotate the dimension to a new position.

Now we are ready to turn this 2-D sketch into a 3-D part with the Extruded Boss/Base command.

Select the Features group of the CommandManager, and the Extruded Boss/Base Tool.

The view of the part changes to display a preview of the extruded solid.

The *base* feature is the first solid feature created. Any subsequent solid features are called *bosses*.

Click and drag the arrow outward to set the direction of the extrusion. When you release the mouse button, a preview of the extruded solid is shown, as in Figure 1.27.

FIGURE 1.27

FIGURE 1.29

On the left side of the screen, where the FeatureManager is normally displayed, the PropertyManager is now active. The PropertyManager allows the properties of the selected entity to be viewed and edited. There are several options available for the extrusion, including adding draft (taper) to the part, but for now we only need to adjust the depth of the extrusion.

Set the depth of the extrusion to 0.75 inches, as shown in Figure 1.28. When you click on the check mark, the circle is extruded into a solid disk, as shown in Figure 1.29.

Now that we have a solid part, we can examine the functions of the viewing tools.

Consider the Standard Views toolbar, which is shown in Figure 1.30. From left to right are the Front, Back, Left, Right, Top, and Bottom Views. The next is the Trimetric View, a good default view for 3-D objects. The Trimetric View differs from an Isometric View in that the front of the model is shown more prominently. The Normal to View can be activated only if a plane, sketch, or planar face is selected. This tool is especially useful if you are working on a sketch that is not oriented parallel to one of the default planes. The final tool opens a dialog box from which these views and some others can be selected. Many users who learned on earlier versions of the SolidWorks program prefer to use the dialog box to select views.

On the View toolbar, shown in Figures 1.31 and 1.32, the first icon allows the previous view to be displayed. Next are the zoom tools, indicated by the magnifying glass icons. The Rotate View Tool is especially useful for visualizing all sides of a model. With this tool active, holding down the left mouse button and moving the mouse causes the model to rotate. The Pan Tool allows the model to be repositioned on the screen without changing the zoom size. (Note: When using the Zoom In/Out, Rotate View, and Pan Tools, it is necessary to turn them off before proceeding. You can do this either by clicking the tool or the esc key.)

The Standard Views Tool allows selection of the same tools contained in the Standard Views toolbar to be selected from a pull-down menu. If desired, you can close the Standard Views toolbar and select views from the pull-down menu (or by clicking the space bar, which causes the

FIGURE 1.28

FIGURE 1.30

FIGURE 1.31

FIGURE 1.32

Standard Views dialog box to be displayed). However, the Standard Views toolbar allows for one-click selection of the desired view.

The appearance of the part can be changed with the display tools. The wireframe tools are sometimes very useful when sketching. Solid views can be displayed either with or without solid lines indicating model edges. The Shadow Tool can be used with either shaded mode to add shadowing to the shaded solid view. A section view can be used to display a cross-section of the part. This is often useful to visualize complex parts. The final tool, the RealView Graphics Tool, allows for a realistic view of the part based on the selected material of the part. This option works only with certain graphics cards, and is not used in this book.

Experiment with these zoom and viewing options. When finished, select a shaded solid display (either with or without edges displayed) and the Trimetric View.

Now we are ready to add to our part. The next feature we will add is the 2.75-inch diameter boss. We could sketch the circle to be extruded on the front or back face of the existing part, but instead we will create a new plane that is 2.25 inches away from the front plane. (There are several reasons why we might want to define the part in this manner. One is that we may want to add draft, a slope to the sides of a feature that allows it to be extracted from a mold. If so, then we want our 2.75-inch dimension to apply at the top of the boss, allowing the diameter to get larger closer to its base.)

FIGURE 1.33

Select the Reference Geometry Tool from the Features group of the CommandManager. From the menu that appears, select Plane (see Figure 1.33).

Select the Front Plane by clicking on it in the FeatureManager, as shown in Figure 1.34 (if the planes are not shown, it is necessary to click the "+" sign next to the part name in order to expand the FeatureManager).

FIGURE 1.34

In the box defining the offset distance, enter 2.25, as shown in Figure 1.35. Click the check mark and the new plane, labeled Plane1, is created.

Of course, the Reverse direction box could have been checked if we wanted the new plane to be behind the Front Plane.

DESIGN INTENT

As we build the 2.75-inch diameter boss of our flange, we can choose from two existing planes/surfaces or construct a new plane. The choice of constructing a new plane in order to allow us to add draft to our part is an example of *design intent*. There are many definitions of design intent. Ours is:

Design intent is the consideration of the end use of a part, and possible changes to the part, when creating a solid model.

Throughout this book, we will identify examples of considering design intent when modeling.

FIGURE 1.35

Now with Plane1 selected (highlighted in green), open a new sketch by selecting the Sketch Tool from the Sketch group of the CommandManager. As shown in Figure 1.36, the sketch grid appears in Plane1.

Sketch   Sketch

Use the Normal to View Tool to change the viewpoint (notice that in this case, the Normal to View is the same as the Front View).

FIGURE 1.36

Select the Circle Tool. Sketch and dimension a 2.75-inch diameter circle, centered at the origin (shown in Figure 1.37 in wireframe mode for clarity).

Circle

Now select the Extruded Boss/Base Tool and switch to the Trimetric View.

As shown by the preview (see Figure 1.38), the extrusion is going away from the base feature rather than toward it. Click and drag on the direction arrow so that the extrusion is in the correct direction, as shown in

Features   Extruded Boss/Base

FIGURE 1.37   **FIGURE 1.38**

Figure 1.39. Since we want the extrusion to extend back to the base feature, we can specify that intent in the definition of the extrusion. In the PropertyManager, select Up To Next from the pull-down menu, as shown in Figure 1.40. Click the check mark to complete the extrusion.

We can turn off the display of Plane1 and of the origin.

From the main menu, select View: Planes, and then select View: Origins.

FIGURE 1.39

FIGURE 1.40

FIGURE 1.41

Both Plane1 and the origin still exist in the model, but turning off the display of planes and origins results in a less cluttered model view, as shown in Figure 1.41.

Now let's add the center hole. This time we will select a face to define our sketch plane. As you move the cursor over the front plane, notice that a flag icon appears. This indicates that a surface will be selected when you click with the left mouse button. Similarly, a line icon indicates that an edge will be selected.

With the flag icon showing, select the front face, as shown in Figure 1.42. It will be highlighted in green.

Select the Normal to View. Select the Circle Tool from the Sketch group of the CommandManager.

Note that selecting a sketch tool (in this case the Circle Tool) with a plane or surface selected causes a sketch to automatically be opened.

Drag out a circle centered at the origin. Select the Smart Dimension Tool and dimension the circle diameter as 1.5 inches, as shown in Figure 1.43.

FIGURE 1.43

FIGURE 1.42

FIGURE 1.44

Now select the Extruded Cut Tool of the Features group of the CommandManager.

Switch to the Trimetric View so that you can see that the direction is correct, and select the type as Through All in the PropertyManager, as shown in Figure 1.44. Click the check mark to complete the cut. The result of this operation is shown in Figure 1.45.

FIGURE 1.45

We will now add the four bolt holes.

Select the surface shown in Figure 1.46. Switch to the Normal to View.

Select the Circle Tool from the Sketch group of the CommandManager, and drag out a circle centered at the origin, as shown in Figure 1.47. In the PropertyManager, check the "For construction" box shown in Figure 1.48.

FIGURE 1.46   **FIGURE 1.47**

FIGURE 1.48

Construction entities help you locate and size sketch parameters, and are indicated by dashed-dotted lines. The circle just drawn represents the bolt circle.

FIGURE 1.49

Select the Smart Dimension Tool. Add a 4.25-inch diameter dimension to the circle, as shown in Figure 1.49.

Select the Centerline Tool, and drag out a vertical centerline from the origin. Make sure that a square appears next to the cursor, indicating that the centerline will begin at the origin, before clicking (see Figure 1.50). Click and drag upward to create the centerline. A "V" next to the cursor indicates that the centerline is vertical (see Figure 1.51).

FIGURE 1.50

FIGURE 1.52

FIGURE 1.54

FIGURE 1.51

The centerline may be difficult to see with the grid display on. Figure 1.52 shows the centerline with the grid display off. Note that all centerlines are construction entities, and the "For construction" box is already checked.

Select the Circle Tool. Drag out a circle from the intersection of the construction circle and the centerline. Note that

the intersection icon appears by the cursor, as shown in Figure 1.53, indicating that the center of the circle will be located precisely at the intersection of the construction

entities. Select the Smart Dimension Tool and add a diameter dimension of 0.50 inches, as shown in Figure 1.54.

The sketch is now fully defined.

Select the Extruded Cut Tool from the Features group of the CommandManager, and extrude a hole with a type of Through All.

The first bolt hole is now in place, as shown in Figure 1.55.

Notice that in the FeatureManager, all of our procedures are being recorded. To more easily identify features for later modifications, we can rename features.

Click once on Cut-Extrude2 to select and highlight the name, as shown in Figure 1.56. Type "Bolt Hole" to rename the feature, as shown in Figure 1.57. Press the Enter key to accept the new name.

FIGURE 1.53

FIGURE 1.55

FIGURE 1.56

FIGURE 1.57

DESIGN INTENT

Why did the circle defining the first bolt hole need to be aligned directly above the origin with the vertical centerline? This step was not absolutely necessary. Imagine that you were given the job of drilling the four bolt holes into a "blank" flange. Where would you drill the first hole? It doesn't matter, as long as the other three holes are equally spaced relative to the first hole.

In this case, leaving the sketch underdefined would have been OK. However, later we will make a 2-D drawing of the flange, and placing the first hole directly above the center of the flange (the origin) in the drawing is the preferred orientation. Since the drawing will be made directly from the part, orienting the bolt holes in the part will make creation of the drawing easier.

FIGURE 1.58

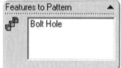

We could create the other three holes in the same manner, but it is easier to copy the single hole into a circular pattern. Also, since our design intent is for the holes to exist in a circular pattern, it makes sense to construct them that way. If we later change the diameter of the holes, the diameter of the bolt circle, or the number of holes, it will be easy to do if we have created them in a pattern.

Make sure that the first bolt hole is selected. From the Features group of the CommandManager, select the Circular Pattern Tool.

Circular Pattern

In the PropertyManager, shown in Figure 1.58, the first box is shaded in pink, indicating that this box is selected for input. This box is where the axis of the circular pattern is to be specified. Although we could create an axis, that is not necessary. Any cylindrical feature of the part has an axis associated with it; we can use one of these axes as the axis for our pattern.

FIGURE 1.59

From the main menu, select View: Temporary Axes, as shown in Figure 1.59.

FIGURE 1.60

Select the axis at the center of the part, as shown in Figure 1.60.

In the Property-Manager, set the number of features to 4, as shown in Figure 1.61.

FIGURE 1.61

FIGURE 1.62

FIGURE 1.63

The "Equal spacing" box should be checked, and the angle set to 360 degrees. The preview of the pattern should now show 4 equally spaced bolt holes, as shown in Figure 1.62. Click the check mark to complete the pattern, which is shown in Figure 1.63.

From the main menu, select View: Temporary Axes to turn off the display of the axes.

Now let's finish the flange by adding *fillets* to three of the sharp edges. A fillet is a feature that rounds off a sharp edge. Actually, a fillet is a rounded edge created by removing material, while a *round* is created by adding material. Fillets and rounds are created by the same command.

From the Features group of the Command-Manager, select the Fillet Tool.

FIGURE 1.66

FIGURE 1.64

FIGURE 1.65

Select the three edges indicated in Figure 1.64 to be filleted. (Be sure to see the line next to the cursor, as shown in Figure 1.65, to indicate that an edge and not a face is being selected.) In the PropertyManager, enter the radius as 0.25 inches, as shown in Figure 1.66. Click the check mark to add the fillets.

Now we can add the *chamfer* to the center hole. A chamfer is a conical feature formed by removing material from an edge.

DESIGN INTENT

The three fillets are added in this tutorial in a single step by selecting the three edges to be filleted within a single fillet command. With this method, only the first fillet is dimensioned. Another way to add the fillets is to close the Fillet Tool after each fillet is created, so that the fillets are created in three separate steps. The preferred method depends on how you wish to edit the fillet radii. If you want all of the fillets to always have the same radius, then the first method allows one value to be changed for all three fillets. If you prefer to edit the fillets separately, then the second method provides an editable dimension for each fillet.

FIGURE 1.67

From the Features group of the CommandManager, select the Chamfer Tool.

Select the inner edge of the center hole, as shown in Figure 1.67. Set the parameters to 0.080 inches and 45 degrees, as shown in Figure 1.68, and click OK to finish. The chamfered part is shown in Figure 1.69.

FIGURE 1.69

Our flange is finished! This part in its present form will be used in Chapter 2 on 2-D drawings, so it needs to be saved.

FIGURE 1.68

From the main menu, select File: Save. Browse to the directory where you want to save the part, and save the part with the file name "Flange." Make sure that the file type is .SLDPRT (part file), as shown in Figure 1.70.

FIGURE 1.70

Leave the part file open for the next section, in which we will learn how to make modifications to the part.

1.3 Modifying the Flange

One of the main advantages of solid modeling is the ability to make changes easily. As we have observed, the FeatureManager has recorded all of the operations required to make the flange, as shown on Figure 1.71. If you click on the plus sign next to each feature, we see that the sketch associated with each feature is stored, as well.

Let's change the first item that we created by increasing the diameter of the base from 5.5 to 7 inches.

FIGURE 1.71

FIGURE 1.72

FIGURE 1.73

FIGURE 1.74

Right-click Sketch1 in the FeatureManager, and select Edit Sketch from the pop-up menu, as shown in Figure 1.72. Double-click the 5.5-inch dimension, and change it to 7.0 inches. When you close the sketch by clicking on the icon in the upper-right corner of the screen (see Figure 1.73), the part will be updated to the new dimension, as shown in Figure 1.74.

An even easier way to edit the sketch dimensions or the extrude depth is illustrated next.

Double-click the icon next to Extrude1 in the FeatureManager.

All of the dimensions used to create the feature are displayed, as shown in Figure 1.75. The sketch dimensions are shown in black, while the feature dimensions (in this case the extrude depth) are shown in blue.

Double-click the diameter dimension and change it back to 5.5 inches. Click the Rebuild Tool.

To add draft to the boss, select Extrude2 from the FeatureManager, right-click and select Edit Feature, as shown in Figure 1.76. In the PropertyManager, turn the draft on (see Figure 1.77) and set the angle to 3 degrees.

FIGURE 1.75

FIGURE 1.76

FIGURE 1.77

FIGURE 1.78

FIGURE 1.79

Check the "Draft Outward" box so that the boss increases in size as it is extruded. Click the check mark to finish.

The draft will be easier to see if you switch to a side view, as in Figure 1.78.

Finally, right-click on CirPattern1 in the FeatureManager and select Edit Feature. Change the number of holes from four to six, as shown in Figure 1.79.

These last two changes illustrate the importance of considering design intent when modeling. If the first boss had been extruded from the base feature, then adding draft would have required us to change the diameter of the boss, calculating the diameter that will result in a 2.75-inch diameter at the top of the boss when draft is included. By sketching in a plane at the top of the boss, the critical 2.75-inch dimension can be maintained easily. Also, by constructing the holes as a circular pattern instead of individually, the number of holes could be modified easily.

Close the part window by clicking on the X in the upper-right corner of the part window, as shown in Figure 1.80. Do not save the changes to the file.

FIGURE 1.80

1.4 An Alternate Method of Modeling the Flange

Here is a different approach to creating the body of the flange. Several sketching features are introduced here.

From the main menu, select File: New, or click on the New File icon. Click on the Part icon and click OK. Click on the Right Plane in the FeatureManger, as shown in Figure 1.81 (see "Design Intent" box on the next page).

Switch to the Right View.

From the Sketch group of the CommandManager, select the Line Tool.

As noted earlier, if a plane or flat surface is selected, then selecting a sketching tool automatically opens a new sketch on that plane or surface.

FIGURE 1.81

DESIGN INTENT

The choice of the initial sketch plane for a part should be considered before beginning a new part. Standard drawing practice is to orient the part so that the Front View is the model view that best describes the part. When we constructed the flange earlier, we selected the Front Plane as the sketch plane for the first feature. The result was the proper orientation of the part, with the Front View being the most descriptive of the standard views.

Front view

Now that we are creating the same part by revolving the part's cross-section about the centerline, we must sketch the section in either the Right or Top Plane to maintain the same part orientation. We have selected the Right Plane.

Although the plane associated with a sketch can be changed by right-clicking the sketch in the Feature-Manager and selecting (Edit Sketch Plane) subsequent operations may be affected, and errors could result. It is good practice to carefully consider the desired orientation of the part before modeling, and selecting the appropriate sketch plane accordingly.

Sketch the profile of horizontal and vertical lines shown in Figure 1.82, with all lines above and to the left of the origin. Make use of the horizontal (Figure 1.83), vertical (Figure 1.84), and alignment (Figure 1.85) drawing aids as you sketch.

FIGURE 1.82

FIGURE 1.83

FIGURE 1.84

FIGURE 1.85

Alignment with
point directly
below

1.156

Centerline

Select the Centerline Tool, and sketch a horizontal centerline from the origin, as shown in Figure 1.86.

Now we will add dimensions to the sketch. Our first dimension will define the diameter of the center hole.

Select the Smart Dimension Tool. Click on the centerline, and then on the horizontal line defining the center hole.

Smart
Dimension

FIGURE 1.86

3.316

FIGURE 1.87

FIGURE 1.88

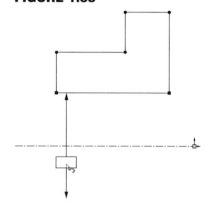

Notice that if you place the dimension above the centerline, a radius dimension is added, as in Figure 1.87. If you drag the dimension below the centerline, a diameter dimension is added, as in Figure 1.88.

Since our part is dimensioned by diameters, we will build the part that way.

Add the 1.5-inch diameter dimension as shown in Figure 1.89.

Add dimension to the other diameters as shown in Figure 1.90.

FIGURE 1.90

FIGURE 1.89

With the Smart Dimension Tool still active, click once on the horizontal line defining the 0.75-inch thickness of the flange. Place a linear dimension. Repeat for the horizontal line defining the 2.25-inch overall height of the flange, as shown in Figure 1.91.

The sketch is still underdefined, as the horizontal position of the sketch relative to the origin is not defined.

Click the esc key to turn off the Dimension Tool. Click on the vertical line of the sketch defining the bottom of the flange, and then, while holding the ctrl key, the origin, as shown in Figure 1.92.

FIGURE 1.91

FIGURE 1.92

(The ctrl key allows you to make multiple selections.)

FIGURE 1.93

In the PropertyManager, add a Coincident relation, as shown in Figure 1.93, which will fully define the sketch.

The fillets can also be added to the sketch. Select the Sketch Fillet Tool. In the PropertyManager, set the radius to 0.25 inches, as shown in Figure 1.94, and then select two edges to be filleted.

Sketch Fillet

The resulting fillet is shown in Figure 1.95.

FIGURE 1.95

FIGURE 1.94

DESIGN INTENT

Relations are used extensively in sketching. In many cases, relations are used instead of dimensions to help to fully define a sketch. Consider the two lines shown here.

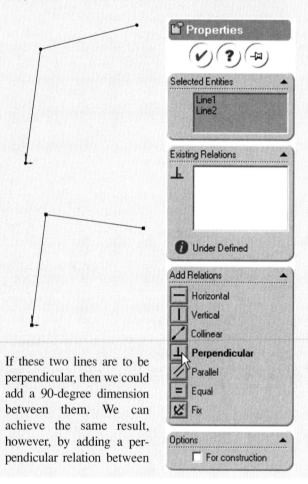

If these two lines are to be perpendicular, then we could add a 90-degree dimension between them. We can achieve the same result, however, by adding a perpendicular relation between

the two lines. To do so, we select one line by clicking on it, and while holding down the ctrl key, selecting the second line. The PropertyManager shows any relations between the selected entities, and provides tools for adding new relations. In this case, the Perpendicular icon is selected, and the lines become perpendicular.

Relations can also be added to individual entities. In these cases, the relations are relative to the predefined coordinate system and origin. For example, if we select

the line extending from the origin, we can add a Vertical relation to the line. Many relations are added as implied relations when the geometry is created. For example, if the line here had been drawn as a vertical line originally, a "V" symbol next to the line would indicate that a vertical relation was being applied. Relations can be deleted if necessary by selecting one of the entities associated with the relation, selecting the

FIGURE 1.96

With the Sketch Fillet Tool still active, select the edges corresponding to the other two fillets. Click on the check mark or press the esc key to end the fillet operation.

The sketch should now appear as in Figure 1.96.

We can also add the chamfer to the sketch. Although there is no Sketch Chamfer Tool on the Sketch Tools toolbar, there is a chamfer command that can be accessed from the main menu under Tools: Sketch Tools: Chamfer. However, we will create the chamfer manually in order to introduce more sketching tools and techniques.

relation in the PropertyManager, and clicking the delete key.

Equal relations are particularly useful for minimizing the number of dimensions required to fully define a sketch. For example, the sketch here will be used to form a link with two holes. Rather than adding a 1-inch dimension to the second circle, an Equal relation can be added by selecting both circles and clicking the Equal icon in the PropertyManager. The diameter of the second circle will then be equal to that of the first circle.

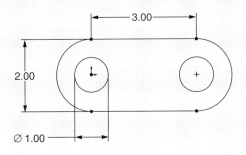

The advantage of this approach is seen if the hole size is to be modified. Changing the 1-inch dimension will cause the diameters of both circles to change.

FIGURE 1.97

Select the Zoom to Area Tool and drag a box around the area where the chamfer is to be added. Select the Line Tool. Begin the line with the lightbulb icon next to the cursor, as shown in Figure 1.97, indicating that the end of the new line will snap to the vertical line.

Drag the line diagonally, ending on the horizontal line (also indicated by the lightbulb icon, as shown in Figure 1.98).

Trim
Entities

Select the Trim Entities Tool from the Sketch group of the CommandManager. Click on the portion of the vertical line

FIGURE 1.98

FIGURE 1.99

FIGURE 1.100

segment to be trimmed away, as shown in Figure 1.99. The line will be trimmed back to the next intersection.

FIGURE 1.101

Repeat for the horizontal line, as shown in Figure 1.100. You will see the message in Figure 1.101 when you try to trim away the end of the line.

Click Yes to perform the trim.

The message relates to the 2.25-inch dimension that we created by dimensioning the length of the horizontal line defining the flange height. Since we have trimmed away part of that line, the dimension defining its length has been lost. We will add that dimension back momentarily.

FIGURE 1.102

Select the Smart Dimension Tool. Add the 0.08-inch vertical dimension between the endpoints of the diagonal line, as shown in Figure 1.102.

Smart Dimension

Click on the diagonal line and then the horizontal or vertical line, and place an angular dimension setting the chamfer angle at 45 degrees, as shown in Figure 1.103.

FIGURE 1.103

Select the Zoom to Fit Tool. With the Smart Dimension Tool still active, click on each of the lines defining the top and bottom of the flange, as shown in Figure 1.104. Add the 2.25-inch height dimension.

The completed sketch is shown in Figure 1.105.

Select the Revolved Boss/Base Tool from the Features group of the CommandManager.

Features

Revolved Boss/Base

FIGURE 1.104

FIGURE 1.105

FIGURE 1.106

By default, the revolution is a full 360 degrees about the sketch's centerline, as shown by the preview of the revolution in Figure 1.106.

Click the check mark in the Property-Manager to complete the revolution.

The resulting solid body is shown in Figure 1.107.

FIGURE 1.107

 Select the surface shown in Figure 1.108. Switch to the Normal to View.

Select the Circle Tool from the Sketch group of the CommandManager, and drag out a circle on the selected surface, in the approximate location shown in Figure 1.109.

 Select the Smart Dimension Tool, and dimension the diameter of the hole as 0.50 inches, as shown in Figure 1.110.

We will now add a second dimension that defines the radius of the bolt circle.

With the Smart Dimension Tool active, click once on the perimeter of the new circle (see

FIGURE 1.109

FIGURE 1.108

Sketch   Circle

FIGURE 1.110

FIGURE 1.111

Figure 1.111) and then click on the origin (see Figure 1.112). **Set the dimension as a radial dimension, as shown in Figure 1.113.**

(Dragging the dimension too far toward one side of the screen or toward the top or bottom of the screen will result in a horizontal or vertical dimension being placed, rather than the desired radial dimension.) Note that the 2.125 dimension can be entered as "4.25/2." The software then performs the math to divide the diameter of the bolt circle by 2.

FIGURE 1.112

FIGURE 1.113

FIGURE 1.114

FIGURE 1.115

FIGURE 1.116

The resulting radial dimension is shown in Figure 1.114. Note that the sketch is still underdefined. This is because the circle that we drew is not tied to any angular location. Since we cannot enter dimensions with values of zero, we specify this location by defining a relation.

Click the esc key to turn off the Smart Dimension Tool. Click on the center point of the new circle, as shown in Figure 1.115, and, while holding down the ctrl key, select the origin, as shown in Figure 1.116.

As noted earlier, the ctrl key allows multiple items to be selected. In the PropertyManager, shown in Figure 1.117, any existing relations between the selected items are displayed. Below the Existing Relations box, there are Add Relation tools. In this case, there are three possible relations between the two points: Horizontal, Vertical, and Coincident. Selecting Coincident will result in an error, since we previously set these points 2.125 inches apart.

Click Vertical to align the two points, and click the check mark to close the PropertyManager.

FIGURE 1.117

FIGURE 1.118

FIGURE 1.119

FIGURE 1.120

The bolt-hole circle is now located directly above the origin, as shown in Figure 1.118. Note that the sketch is now fully defined.

From the main menu, select Tools: Sketch Tools: Circular Step and Repeat, as shown in Figure 1.119. Click on the circle to identify it as the item to be repeated.

The center of the pattern will be x = 0 and y = 0 (the origin), and the number of steps = 4, as shown in Figure 1.120. Click OK to complete the pattern.

The completed sketch is shown in Figure 1.121.

Select the Extruded Cut Tool from the Features group of the CommandManager. Set the type as Through All to create the holes, as shown in Figure 1.122.

Features   Extruded Cut

The volume and mass of the flange can be calculated. The first step in this process is to identify the material of the part. The program has a materials library containing the properties of many common materials.

FIGURE 1.121

FIGURE 1.122

FIGURE 1.123

FIGURE 1.126

FIGURE 1.124

In the FeatureManager, right-click on Material, and select Edit Material, as shown in Figure 1.123. In the Materials Editor that appears, select 6061 Aluminum Alloy, as shown in Figure 1.124.

Note: The density of the material can be displayed by clicking the Create/Edit Material button and then the Physical Properties tab of the pop-up box. For 6061 aluminum, the density is 2700 kg/m³. Even though the unit system of the part is inches, pounds, and seconds, the material library contains properties that have been entered in SI units. However, when the Mass Properties Tool is used, the properties are expressed in the units selected for the part. In this case, the mass of the flange will be displayed in pounds. (To be absolutely correct, we should say that the mass of the flange is 2.14 pounds-mass or that the *weight* of the flange is 2.14 pounds.)

From the main menu, select Tools: Mass Properties.

In addition to the volume of 21.93 cubic inches and mass of 2.14 pounds, the program locates the center of mass (0.68 inches

FIGURE 1.125

Mass, Volume, and Surface Area

Coordinates of Center of Mass

Location of Center of Mass

Dynamics (Kinetics)

In physics, you learned Newton's First Law:

$$F = ma$$

or the forces acting on a body equal the mass of the body times its acceleration. For mechanical engineers designing machines with moving components, application of this law results in the forces required to move a body in a particular manner. In the field of engineering materials, the development of lightweight, strong materials allows moving components to be moved at extremely high accelerations, resulting in higher performance.

The form of Newton's First Law written above applies to bodies moving in linear or translational motions. Most machine components also move in rotational motions. Newton's First Law for rotational motions is:

$$T = I\alpha$$

or the torques (T) acting on a body equal the mass moment of inertia (I) times the angular acceleration (α) of the body. This mass moment of inertia is a function of the body's mass and the way in which the mass is distributed relative to the axis of rotation. For example, consider the two wheels shown here. The two wheels are the same diameter and have the same weight, but the wheel on the left has a mass moment of inertia almost

twice that of the wheel on the right. If the wheels were mounted onto shafts, it would take twice as much torque to bring the wheel on the left up to speed as it would the wheel on the right. While a low mass moment of inertia is usually desirable, a notable exception is a *flywheel*. The inertia of a flywheel in an engine rotating at a high speed makes it difficult to slow down. As a result, the rotational speed of the engine remains smooth despite small fluctuations in power.

The calculation of mass and mass moments of inertia for complex shapes can be tedious. Therefore, the use of solid modeling software to perform these calculations saves time for engineers performing these types of analyses.

---

from the back face of the flange), and calculates surface area and moments of inertia, as shown in Figure 1.125 (see Future Study box).

Finally, we can give our flange a realistic look by applying a *texture* to the surfaces.

 Select the entire part by clicking on the part name in the FeatureManager. Select the Textures Tool from the Standard toolbar. In the Texture Manager, select Metal: Cast: Cast Rough, as shown in Figure 1.126. Use the slider bar to adjust the grain size as desired.

Textures can also be applied to individual surfaces. For example, a part might be cast with certain critical surfaces machined.

The completed model is shown in Figure 1.127.

FIGURE 1.127

DESIGN INTENT

The main flange body has been modeled in two different ways: with a series of extrusion, fillet, and chamfer operations, and with a single revolution. Which is the better method? Although the final models are identical, editing and troubleshooting are more difficult with the second method. Many errors encountered when modeling parts are associated with sketches: duplicate entities, open contours, endpoints that do not meet, and overdefined sketches are all common mistakes that can be difficult to detect and correct in complex sketches. The best approach is simplicity: use many simple steps rather than a few complex steps. Following this approach will result in models that are easier to edit and debug.

Problems

P1.1   Create a solid model of the stepped shaft, using Extruded Boss/Base and Fillet features. Dimensions shown are inches.

Dimensions are in inches

P1.2   Use an Extruded Boss to create a solid model of a 2-foot long segment of a W8×31 steel wide-flange beam (The designation W8×31 denotes wide-flange beam family, 8 inches nominal depth, weight of 31 pounds per foot.) Set the material type to plain carbon steel (density = 7800 kg/m³), and find the weight of the part in pounds. Dimensions shown are inches.

Use Equal relations so that the part cross-section is modeled using only the dimensions shown. Use a fully defined sketch of the cross-section.

P1.3   Create a solid model of the pulley. Create the cross-section, revolve it into a solid, and then add the keyway. Set the material to cast carbon steel (density = 7800 kg/m³) and find the weight of the pulley. Dimensions shown are inches.

Use Equal relations so that the part cross-section is modeled using only the dimensions shown. Use a fully defined sketch of the cross-section.
(Answer: Weight = 8.26 lb)

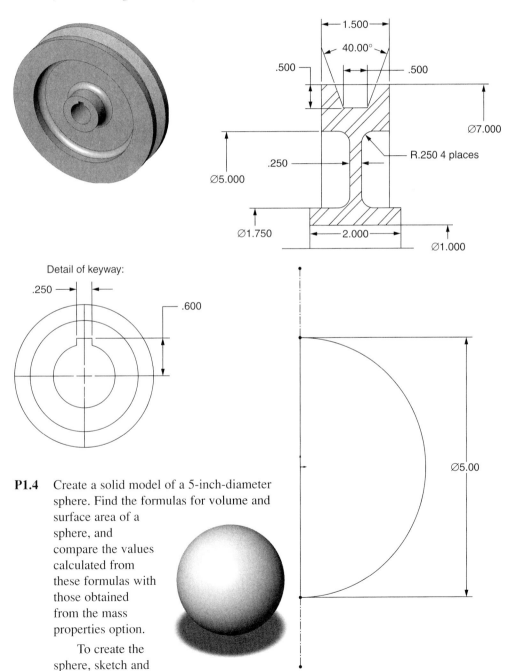

Detail of keyway:

P1.4   Create a solid model of a 5-inch-diameter sphere. Find the formulas for volume and surface area of a sphere, and compare the values calculated from these formulas with those obtained from the mass properties option.

To create the sphere, sketch and

dimension a circle, and pass a centerline through the center of the circle. Trim away half of the circle. Close the sketch geometry by adding a line connecting the ends of the semicircle. Then revolve the sketch 360 degrees about the centerline. Note: Be sure that you have only one centerline in the sketch. Multiple centerlines cause the Revolve command to fail.

P1.5   Create a solid model of this plastic pipe tee. A tee is used to connect pipes together. The type of tee shown here is used to join pipes with solvent welding. A chemical is applied to the inside of the socket, and the pipe is then forced into the socket. The solvent softens the plastic, and when the solvent dries, a strong, permanent joint is created. The sockets are tapered slightly to allow for a tight fit with the pipe.

Section A-A

Sockets are to be tapered 0.50 degrees

Set the material to PVC Rigid (density = 1300 kg/m^3) and find the weight of the tee. Dimensions shown are inches.
(Answer: Weight = 0.1044 pounds. Click the Options tab in the Mass Properties box and increase the number of decimal places, if necessary.)

P1.6   Create a solid model of this 2-mm-thick steel bicycle disk brake rotor. Calculate the mass of the part, using AISI 304 stainless steel as the material (density = 8000 kg/m^3). Dimensions shown are mm.

P1.7   Create a solid model of the pleated filter element. Dimensions shown are inches, and the diameter dimensions shown are *nominal* dimensions. There are 36 pleats in the part, and the part is 8 inches long.

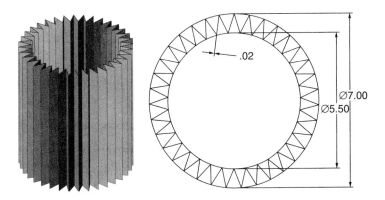

    To create this part, start a sketch with two construction circles representing the nominal inner and outer diameters. Add and dimension two lines representing one pleat, and use a Circular Step and Repeat to copy these two lines into the other 35 positions. Select the Extruded Boss Tool. Since the sketch is an open contour, a thin-feature extrusion will be created. Set the thickness to 0.02 inches, and set the type as Mid-Plane.

Engineering Drawings

Introduction

Even companies that design parts exclusively with solid modeling software still need to produce 2-D drawings. A multiple-view 2-D drawing is the best way to document the design, showing all of the dimensions necessary to produce the part. Since no manufacturing process can produce "perfect" parts, the *tolerances* (allowed variations from the stated dimension of a part) allowed for important dimensions are also shown on 2-D drawings.

The SolidWorks program allows for 2-D drawings to be quickly and easily produced from 3-D models. The drawing will be fully associative with the part file, so that changes to the part will be automatically reflected in the drawing, and vice versa.

In this chapter, you will:

- make a 2-D drawing from a SolidWorks part,
- create a custom drawing sheet format, and
- use eDrawings™ software to create a drawing file that allows for easy file sharing and collaborative editing.[1]

2.1 Drawing Tutorial

In this tutorial, a 2-D drawing of the flange from Chapter 1 will be made.

Open the SolidWorks program. Click on the New Document Tool.

Select the Drawing icon, as shown in Figure 2.1, and click OK.

FIGURE 2.1

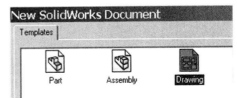

As if you were making a drawing by hand, your first step is to select the size and type of paper to be used. Later we will look at using sheet formats with a title block, but for now let's use a blank $8\frac{1}{2}" \times 11"$ sheet (A-Landscape).

Choose a paper size of A-Landscape, and click on the box labeled "Display sheet format" to clear the check mark, as shown in Figure 2.2. Click OK to close the dialog box.

If the Model View box appears in the PropertyManager, as shown in Figure 2.3, click on the box labeled "Start command when creating new drawing" to clear the check mark, and click on the × to close the box.

---

[1] eDrawings is a trademark of the SolidWorks Corporation, 300 Baker Avenue, Concord, MA 01742.

FIGURE 2.2

FIGURE 2.3

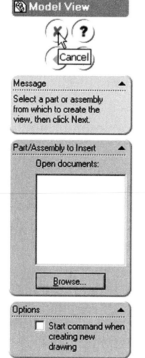

Before bringing in the flange model data, we will set a few options.

Select Tools: Options from the menu. Under the System Options tab, select Drawings: Display Style, and set the hidden line and tangent displays as shown in Figure 2.4

Whether or not to show hidden and tangent edges is sometimes dependent on the complexity of the part. As we will see later, we can change these display options for each model view.

Under the Document Properties tab, select Detailing and set the dimensioning standard to ANSI, as shown in Figure 2.5.

Select Units, and set the Unit system to IPS (inch, pound, second), and the decimal places to 2, as shown in Figure 2.6. Click OK.

FIGURE 2.4

FIGURE 2.5

Although the part was modeled with the decimal places set to 3, the number of decimals does not affect the accuracy of the model, only the way dimensions are displayed. For a drawing, the number of decimal places should be related to the tolerance level of most of the dimensions.

In the drawing space, right-click and select Properties from the menu that appears, as shown in Figure 2.7. Set the Type of Projection to Third Angle. Click OK to close the Sheet Properties box.

A third angle projection sets the front view in the lower-left corner, while a first angle project sets the front view in the upper-right corner, as shown in Figure 2.8.

Don't worry about the drawing scale. It will automatically adjust when we import the part, and we can change the scale at any time by pulling up the Sheet Properties menu. We can also change the paper size and format at any time.

FIGURE 2.6

FIGURE 2.7

FIGURE 2.8

When we modeled a part in Chapter 1, we found most of the commands we needed in the CommandManager. In the drawing mode, the CommandManager has three groups of commands: Drawing, as shown in Figure 2.9, Sketch, as shown in Figure 2.10, and Annotation, as shown in Figure 2.11. The group is changed by clicking on the appropriate icon at the left of the CommandManager.

FIGURE 2.9

FIGURE 2.10

FIGURE 2.11

We are now ready to import the geometry from the part file.

From the Drawing group of the CommandManager, select the Standard 3 View Tool.

If you have the flange open in another window, then it will appear in the Property-Manager, as shown in Figure 2.12. If not, then click the Browse... button and find the flange part file where you stored it.

FIGURE 2.12   **FIGURE 2.13**

When you select the part file, three standard views are inserted into the drawing, as shown in Figure 2.13.

If a dialog box appears asking you to specify the tangent edge display, select "Don't ask me again" and close the box.

As you pass the cursor over each drawing view, notice that a rectangular box that defines the boundary of that view appears. Moving the cursor over the drawing view boundary displays the move arrows, as shown in Figure 2.14. When the move arrows are displayed, you can click and drag the drawing view to a new location. Note that the views remain in alignment as they are moved.

The right and top views are essentially identical, so only one is required to communicate the design parameters.

Click once on the right view to select it (its boundary will be displayed in green), and press the delete key. A dialog box appears, as shown in Figure 2.15, requesting confirmation of the delete command. Click Yes to delete the view.

Now we will import the dimensions that were used to create the model.

FIGURE 2.15

FIGURE 2.14

FIGURE 2.16

FIGURE 2.17

FIGURE 2.18

Select the Front View and select Insert: Model Items from the main menu. Note that Dimensions is selected, as shown in Figure 2.16. Click OK, and all of the dimensions that can be shown in this view will appear, as shown in Figure 2.17.

It may be necessary to zoom out for you to see all of the dimensions.

Select the Top View and select Insert: Model Items to apply dimensions to that view, as shown in Figure 2.18.

If you want to change the font of the dimensions, select Tools: Options from the menu. Under Document Properties, select Annotations Font: Dimensions, and select a font and font size from the dialog box shown in Figure 2.19. The font size may be input in inches or points.

The placement of dimensions can be modified by clicking and dragging on the dimension value, as illustrated in Figure 2.20.

The arrow heads can be displayed inside or outside the dimension lines by clicking on the green dot on one of the arrow heads, as illustrated in Figure 2.21.

For diameter dimensions, the angular placement of the dimension can be changed by dragging the green dot above the dimension, as illustrated in Figure 2.22.

In the front view, the bolt circle is difficult to distinguish from the tangent edges that are displayed. For this view, hiding the tangent edges may be desired.

DESIGN INTENT

In Chapter 1, we learned how to orient the model so that the front view was the view that best describes the part. Similarly, more dimensions will be applied to the front view than to the other views. For that reason, we applied dimensions to the front view first. Dimensions that could be applied to either the front or the top view, such as the 5.5-inch diameter, will be applied only to the front view. It is good practice to always insert dimensions to the front view first. Dimensions can be moved from one view to another, as will be demonstrated later in the tutorial.

FIGURE 2.19

FIGURE 2.20

FIGURE 2.21   **FIGURE 2.22**

Right-click within the front view boundary, and select Tangent Edge: Tangent Edges Removed, as shown in Figure 2.23.

The front view should now appear as in Figure 2.24. Note that the tangent edges are still visible in the top view.

A section view will help show the center hole and the chamfer more clearly

FIGURE 2.23   **FIGURE 2.24**

DESIGN INTENT

In Chapter 1, we defined the position of the bolt holes in two ways: first by adding a construction circle and adding a diameter dimension to that circle, and second by adding a radial dimension from the center. At the time, there seemed to be no real advantage to either method. When dimensioning the drawing, however, we see that drawing a bolt circle and dimensioning its diameter is the preferred way to show the radial position of the bolt holes. If the distance were defined by a radial dimension,

then we could simply delete the radial dimension in the drawing, add a construction circle, and add a new dimension. However, the new dimension would not be associative with the part. That is, changing the added 4.25-inch dimension will not change the bolt hole positions, and changing the radial dimension in the part would not cause the drawing dimension to be updated. Therefore, maintaining full associativity between the part and the drawing would require editing of the part file.

Select the Section View Tool from the Drawings group of the CommandManager.

The line tool becomes active when the Section View Tool is selected, as shown by the line icon next to the cursor (see Figure 2.25). Since we want to draw our line exactly through the center of the part, we need to "wake up" a center mark for placement of the line's endpoints.

FIGURE 2.25

Without clicking any mouse button, move the cursor over the edge of one of the circular features, as shown in Figure 2.26, and hold it there momentarily. Move the cur-

FIGURE 2.26

FIGURE 2.27

FIGURE 2.28

FIGURE 2.29

sor outside of the edge of the flange. When the dotted line appears that indicates alignment with the center of the flange, as in Figure 2.27, click and drag a horizontal line completely through the part, as shown in Figure 2.28.

When you finish the line, a section view is created. Move the cursor to place the section view. By default, the view will be kept in alignment with the front view. Drag the section view well away from the other views, even if you need to drag it completely out of the drawing borders, and click once to place the view, as shown in Figure 2.29.

Right-click within the Section View boundary box, and from the menu that appears, select Alignment: Break Alignment, as shown in Figure 2.30. Move the cursor over the edge of the Section View boundary box so that the move arrows appear, as shown in Figure 2.31. Click and drag the view to a new location on the sheet.

When we created the section view, the direction of the view relative to the section line was down, toward the bottom of the sheet. This resulted in a section view that is upside-down from our preferred orientation of the part. This is easily modified.

FIGURE 2.30

FIGURE 2.31

FIGURE 2.32

Double-click on the section line in the front view. The arrows defining the direction of the section will be reversed, as shown in Figure 2.32.

The section view will not be immediately reversed. Instead, it appears crosshatched, as shown in Figure 2.33, indicating that it needs to be rebuilt.

Click the Rebuild Tool in the Standard toolbar.

The section view appearance is now consistent with the direction of the section arrows, as shown in Figure 2.34.

FIGURE 2.33

Section A-A

FIGURE 2.34

Section A-A

With the Section View selected (highlighted with a green boundary), select the Centerline Tool from the Annotations group of the CommandManager.

FIGURE 2.35

Section A-A

Centerlines will be automatically added to the holes, as shown in Figure 2.35. Close the Centerline Tool by clicking on the check mark in the PropertyManager, as shown in Figure 2.36.

Dimensions can be imported into section views. In this case, the 45-degree angle of the chamfer can be seen only in the section view.

FIGURE 2.36

Click in the white space of the Section View boundary box to select this view, and from the main menu, select Insert: Model Items. Click OK to import dimensions into the view. The 45-degree dimension will be added, as shown in Figure 2.37.

The chamfer is difficult to see at the default scale. (The scale of each view is shown in the PropertyManager as each view is selected. For this drawing, the default scale is 1:2.) A detail view that enlarges the chamfer region will be helpful.

Select the Detail View Tool from the Drawings group of the CommandManager. The Circle Tool will be activated automatically.

FIGURE 2.37

Drag out a circle around the area to be included in the detail view, as shown in Figure 2.38.

FIGURE 2.38

FIGURE 2.39

FIGURE 2.40

Move the cursor to the position where you want the detail view to appear, and click to place the view, as shown in Figure 2.39.

Note that the detail view has a scale of 1:1. The scale of any drawing view can be changed, and we will change the scale of the detail view to enlarge the chamfer area further.

FIGURE 2.41

With the Detail View selected, check the "Use custom scale" box in the PropertyManager, as shown in Figure 2.40. Change the scale to 2:1, and click the check mark.

You can change the location of the detail view by clicking and dragging on the center of the Detail circle in the section view (note the Detail Circle icon next to the cursor in Figure 2.41).

FIGURE 2.42

Section A-A

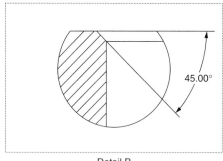

Detail B
Scale 2:1

At this point, we would like to move some of the dimensions from one view to another. We could simply add new dimensions where we want them and delete them from the other view, but doing so would cause the associativity with the part to be lost for that dimension.

Press and hold the shift key. Click and drag the 45-degree chamfer dimension from the section view into the Detail View boundary box. It is not necessary to place the dimension exactly where it belongs. When you release the mouse button, the dimensions will be moved into the detail view. Release the shift key, and drag the dimension to its final position, as shown in Figure 2.42.

Repeat for the 0.08-inch chamfer dimension, and also move the 1.50-inch diameter dimension to the section view. Your drawing should appear similar to Figure 2.43.

As we noted earlier, the number of decimal places displayed in the drawing will relate to the tolerances applied to that dimension. Often, certain dimensions are more critical than others and will need to have tolerance values specified. For example, suppose that the center 1.5-inch diameter hole needs to have a tight tolerance to allow for a good fit with the part that will be inserted into the hole. When that dimension is selected, a number of options appear in the PropertyManager.

FIGURE 2.43

FIGURE 2.44

Click on the number portion of the 1.50-inch dimension. In the PropertyManager, select a tolerance type of Bilateral from the drop-down menu, as shown in Figure 2.44. Enter a plus value of .003 and a minus value of 0. In the Precision Boxes for the dimension and the tolerance (the boxes that have .12 displayed), use the pull-down menus to select .123 (3 decimal places). Click the check mark to apply the format to the dimension.

The dimension should now appear as in Figure 2.45, indicating that the diameter can be no smaller than the 1.500-inch nominal dimension, but can be up to 0.003 inches larger. It may be necessary to click and drag the Section A-A label to allow room for the dimension with tolerances.

Since we are not using a title block with default tolerances shown, we will specify them in a note.

Select the Note Tool from the Annotation group of the CommandManager. Drag the cursor to the approximate location where the

FIGURE 2.45

Section A-A

Manufacturing Processes, Dimensioning and Tolerancing, and Metrology

In this example, we placed a relatively tight tolerance on the diameter of the center hole so that a tight fit could be realized between the flange and the part that fits into the hole. How do you know what the tolerance should be? Perhaps a better question is: what tolerance is *required* for the part to function as designed in an assembly? The answer to this question will in many cases dictate the manufacturing process that can be used. Many engineering students study manufacturing processes, but it is impossible to learn the specific details of all of the manufacturing processes in use today. On the job, engineers should learn as much as possible about the manufacturing processes used by their company and its suppliers, in order to understand what tolerances are practical and economical.

Applying dimensions and tolerances to engineering drawings is a more complex topic than it appears. In addition to the tolerance values applied to linear dimen-

sions, there are many other geometric parameters, such as flatness, straightness, perpendicularity, and surface finish, that may need to be specified for many components. At many companies, the detailing of drawings is performed by a drafting department, and engineers document designs with less-formal drawings and sketches. At many smaller companies, however, engineers detail their own drawings. For these engineers, study of dimensioning and tolerancing standards and practices is necessary.

A related area of study is *metrology*, the science of measurements. Many features are difficult to measure accurately with traditional methods, and computer-controlled Coordinate Measuring Machines (CMMs) are now widely used in industry. Using statistical methods with measured data, quality assurance engineers track variations and work to control the processes that are used to produce the components. This method of *Statistical Process Control* (SPC) allows problems to be detected and corrected before reject parts are produced.

FIGURE 2.46

All dimensions are inches and degrees
Default tolerance for linear dimensions is ± .020
Default tolerance for angular dimension is ± 2|

Annotati...   Note

note will be placed, and click. In the box that appears, begin typing the text of the notes shown in Figure 2.46.

Use the Enter key to start a new line. The ± symbol is inserted by clicking on the Add Symbol Tool in the PropertyManager and selecting Plus/Minus from the list of available symbols, as shown in Figure 2.47.

FIGURE 2.47

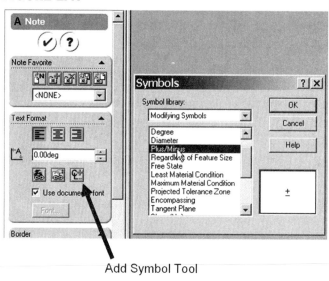

Add Symbol Tool

FIGURE 2.48

Click outside of the text box to place the note. Press the esc key to turn off the Note Tool; otherwise, subsequent mouse clicks will place duplicate notes on the drawing.

If desired, you can change the font type and size by clicking on the note to select it and unchecking Use Document's Font in the PropertyManager and then selecting the Font button, as shown in Figure 2.48. A new font or font size can then be chosen.

Since we defined all of the fillets with a single fillet command in the part, only one of the fillets is dimensioned in the drawing. We might prefer to call out the fillet radius in a note, yet maintain associativity with the part.

FIGURE 2.49

Double-click on the notes. Move the cursor to the end of the last line and press the Enter key to move to a new line. Type in "All fillets and rounds are to be." Click on the .25-inch dimension, as shown in Figure 2.49. The dimension's value will appear in the text box. Add "radius."

FIGURE 2.50

All dimensions are inches and degrees
Default tolerance for linear dimensions is ± .020
Default tolerance for angular dimension is ± 2
All fillets and rounds are to be .25 radius

FIGURE 2.51

to complete the note, and click outside of the text box to end editing of the note. The completed note is shown in Figure 2.50.

FIGURE 2.52

If you change the value of the radius, then the value in the note will change accordingly. Now that the fillet radius is shown in a note, displaying it on the drawing is unnecessary. However, deleting the dimension will produce an error, since the note links to the value of the dimension. Therefore, we will hide the dimension.

Right-click on the R.25 dimension and select Hide from the menu, as shown in Figure 2.51.

If you want to show the dimension later, you can do so by selecting the Hide/Show Annotations Tool from the Drawings group

FIGURE 2.53

of the CommandManager. Any hidden dimensions will show up light, and clicking on them will change them back to visible dimensions, as shown in Figure 2.52. Similarly, clicking on visible dimensions will hide them. This is good method for hiding several dimensions at once. When finished, select the Hide/Show Annotations Tool again to return to the normal editing mode.

A trimetric view can often be helpful in interpreting a 2-D drawing.

Select the Model View Tool from the Drawing group of the Command-Manager. The part can be selected by either double-clicking the part name in the PropertyManager, as shown in Figure 2.53, or by clicking in any of the existing drawing views. In the PropertyManager, select the Trimetric View as the orientation of the new model view, as shown in Figure 2.54.

FIGURE 2.54

FIGURE 2.55

All dimensions are inches and degrees
Default tolerance for linear dimensions
Default tolerance for angular dimensions

Move the cursor to the desired location of the new view, and click to place the view, as shown in Figure 2.55.

 If desired, select a shaded mode for this drawing view.

 Move the drawing views and notes around the sheet so that they can all be seen clearly. Add a note identifying the scale as 1:2. To see how the drawing will look when printed, select the Print Preview Tool.

Your drawing should appear similar to Figure 2.56.

Select File: Save and save the drawing as "Flange," with a file type of .SLDDRW (SolidWorks Drawing).

Open the Flange part file, and experiment with changing dimensions and noticing the updating of the drawing. Similarly, change some drawing dimensions and observe the changes in the part.

FIGURE 2.56

Section A-A

Detail B
Scale 2:1

Scale = 1:2 unless noted
All dimensions are inches and degrees
Default tolerance for linear dimensions is ± .020
Default tolerance for angular dimension is ± 2
All fillets and rounds are to be .25 radius

FIGURE 2.57

FIGURE 2.58

FIGURE 2.59

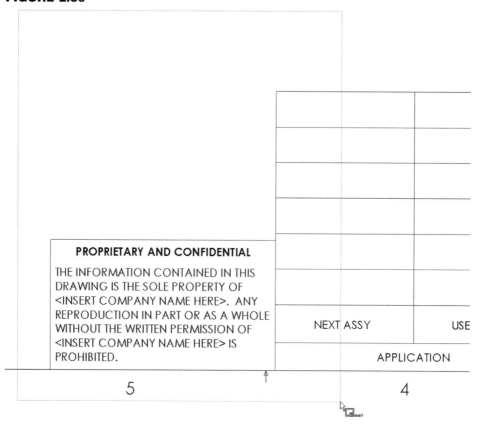

2.2 Creating a Drawing Sheet Format

In this section, we will create a simple drawing sheet format.

Select File: New: Drawing from the main menu. Select a standard sheet format, the A-Landscape format, from the pop-up menu.

Right-click in the drawing area and from the menu that appears, select Edit Sheet Format, as shown in Figure 2.57.

This will allow you to select and modify features associated with the format. The default sheet format is shown in Figure 2.58. Items can be added and deleted from this default format. For our purposes, we will create a very simple sheet format with only a few items.

Click and drag a box around the proprietary note, as shown in Figure 2.59, and press the delete key.

Sketch   Trim Entities

Continue to select and delete the Applications block. Where necessary, use the Trim Entities Tool from the Sketch group of the CommandManager to trim away portions of lines, as shown in Figure 2.60.

FIGURE 2.60

^ title block
add a part
scale and sheet

be filled in for each
the scale and sheet num-

To add a company name,
select the Note Tool from
the Annotation group of the
CommandManager. When
the note, do not click on any entity,
be added to the note, as shown in
add notes specifying a company name,
initials to the "Drawn" box, as shown in
If necessary, change the font sizes so that
fit the boxes.

	UNLES
	DIMEN
	TOLER/
	FRACTI
	ANGUI
	TWO P
	THREE
	INTERPR
	TOLERA
	MATERI/
	FINISH
	DO

4

2.61

	NAME	DATE	
			TITLE:
ERWISE SPECIFIED:			
ARE IN INCHES		SIZE	DWG. NO.		REV
DECIMAL ±		A			
E DECIMAL ±		SCALE: 1:2		SHEET 1 OF 1
E DRAWING				

way to change the font type, size, and orientation is from the Font toolbar,
wn in Figure 2.64. If the Font toolbar is not visible, right-click on any of the

.62

toolbars and click on Font in the list of
available toolbars.

When the format is the way you want it,
select File: Save Sheet Format, as shown
in Figure 2.65.

Give your sheet format a unique name,
as shown in Figure 2.66, and click Save.

FIGURE 2.63

	NAME	DATE
DRAWN	EH	
CHECKED		
ENG APPR.		
MFG APPR.		
Q.A.		

UNLESS OTHERWISE SPECIFIED:

DIMENSIONS ARE IN INCHES
TOLERANCES:
ANGULAR: ±
TWO PLACE DECIMAL ±
THREE PLACE DECIMAL ±
DO NOT SCALE DRAWING

Milwaukee School of Engi

TITLE:

SIZE	DWG. NO.		R
A			
SCALE: 1:2		SHEET 1 C	

FIGURE 2.64

Font   ✕

[A] Century Gothic ▾ 24 ▾ 0.25in **B** *I* U̲ ≣ ≣ ≣

FIGURE 2.65

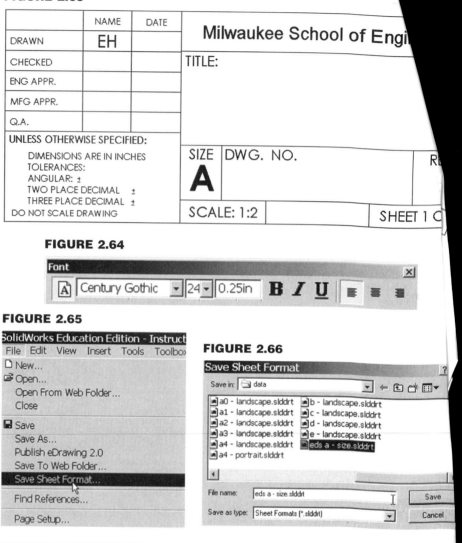

SolidWorks Education Edition - Instruct
File Edit View Insert Tools Toolbox

🗋 New...
📂 Open...
   Open From Web Folder...
   Close

💾 Save
   Save As...
   Publish eDrawing 2.0
   Save To Web Folder...
   Save Sheet Format...
   Find References...
   Page Setup...

FIGURE 2.66

Save Sheet Format

Save in: 🗀 data   ▾ ← 🖭 📷 ▦ ▾

🔊 a0 - landscape.slddrt   🔊 b - landscape.slddrt
🔊 a1 - landscape.slddrt   🔊 c - landscape.slddrt
🔊 a2 - landscape.slddrt   🔊 d - landscape.slddrt
🔊 a3 - landscape.slddrt   🔊 e - landscape.slddrt
🔊 a4 - landscape.slddrt   🔊 eds a - size.slddrt
🔊 a4 - portrait.slddrt

File name: eds a - size.slddrt   Save

Save as type: Sheet Formats (*.slddrt) ▾   Cancel

FIGURE 2.67

Sheet Properties   ? ✕

Name: Sheet1   Type of projection
                    ○ First angle   Next view label: H
Scale: 1 : 2   ● Third angle   Next datum label: A

Sheet Format/Size
● Standard sheet size   Preview:
┌─────────────────────┐ [Reload]
│ A - Landscape   ▲│
│ A - Portrait   │
│ B - Landscape   │
│ C - Landscape   │
│ D - Landscape   │
│ E - Landscape   │
│ A0 - Landscape   ▼│
└─────────────────────┘
E:\PROGRAM FILES\SW200 [Browse...]
☑ Display sheet format

○ Custom sheet size   Width: 11.00in   Height: 8.50in
  Width: Height:

Use custom property values from model shown in:
Default   ▾   OK   Cancel

Close the drawing window. When asked if you want t
save the drawing file, click No.

The drawing sheet format can be selected when begin
ning a new drawing, and can be applied to existing draw
ings. We will now apply the format to the flange drawing.

Open the flange drawing saved earlier. Right-click in
the drawing area (outside of any view's boundary
box) and select Properties from the menu. In the
Sheet Properties box that appears, click on the
"Standard sheet size" button, then choose the Browse
key and find the sheet format just saved. Check the
"Display sheet format" box, as shown in Figure 2.67,
and click OK.

FIGURE 2.68

FIGURE 2.69

.75

2.25

B

Ø1.500 $^{+.003}_{-.000}$ Section A-A

Ø5.50

Ø2.75

Ø.50

.08

45.00°

A   A

Ø4.25

Detail B
Scale 2:1

	NAME	DATE	Milwaukee School of Engineering	

Scale = 1:2 unless noted
All dimensions are inches and degrees
Default tolerance for linear dimensions is ± .020
Default tolerance for angular dimension is ± 2
All fillets and rounds are to be .25 radius

A   REV

SCALE: 1:2   SHEET 1 OF 1

The sheet format will be applied to the drawing, as shown in Figure 2.68.

Right-click in the drawing area and select Edit Sheet Format, as shown in Figure 2.69.

Add notes for the part name and number and the date. Edit the tolerance data, as shown in Figure 2.70.

FIGURE 2.70

	NAME	DATE	Milwaukee School of Engineering		
DRAWN	EH	7 July 04			
CHECKED			TITLE:		
ENG APPR.			**FLANGE**		
MFG APPR.					
Q.A.					
UNLESS OTHERWISE SPECIFIED:					
DIMENSIONS ARE IN INCHES TOLERANCES: ANGULAR: ± 2 TWO PLACE DECIMAL ± .020 THREE PLACE DECIMAL ± .005 DO NOT SCALE DRAWING			SIZE **A**	DWG. NO. A04001	REV
			SCALE: 1:2		SHEET 1 OF 1

FIGURE 2.71

Right-click in the drawing area, and select Edit Sheet, as shown in Figure 2.71.

Note that the drawing itself and the title block information cannot be edited at the same time. You must toggle between Edit Sheet, to edit the drawing information, and Edit Sheet Format, to edit the title block and any other features contained in the sheet format definition.

Move around the drawing views so that they fit within the sheet boundaries and around the title block. Edit the notes to remove the scale and tolerance values, since the title block contains that information.

The finished drawing is shown in Figure 2.72.

Save this drawing for use in the eDrawings tutorial that follows.

FIGURE 2.72

2.3 Creating an eDrawing

One disadvantage of using drawings that are fully associative with part files is that in order to share the file with other users, they must have access to both the drawing file and the associated part file. The SolidWorks program does have an option called "RapidDraft," which creates drawings that can be loaded without the part file. This option can be useful for organizations in which several persons need to work on the drawing and part simultaneously. For example, a draftsman can work on detailing and annotating a drawing while an analyst modifies the part file to reduce stresses in the part. Later, the part and drawing can be

FIGURE 2.73

synchronized, and the drawing will be updated to reflect the changes to the part file.

To share drawings with others, including those who do not have access to SolidWorks software, eDrawings are convenient. The eDrawings Viewer can be downloaded free, and the eDrawings Publisher also allows the creation of executable files that can be read on any computer. Even within organizations using SolidWorks software, eDrawings are extensively used to share information. Markup tools allow users to make comments and corrections, and the small file sizes of eDrawings make exchange of data via e-mail possible.

To save a SolidWorks drawing as an eDrawing, simply select File: Save As from the main menu and select eDrawing (.edrw) as the file type. The eDrawing file format is very efficient, and file sizes produced are small. For example, the file sizes of the part and drawing files of the flange in Chapters 1 and 2 are:

- Part file: 591 kB
- SolidWorks drawing file: 461 kB
- eDrawings file: 42 kB

To create an executable file or to mark up an eDrawing, it is necessary to have the eDrawings program installed on your computer.

FIGURE 2.74

Open the flange drawing file that you saved in the previous section.

From the main menu, select Tools: Add-Ins, as shown in Figure 2.73. If the box next to eDrawings is not checked, click on the box to activate eDrawings, as shown in Figure 2.74. Click OK.

 If the eDrawings toolbar is not shown, right-click on any toolbar and select eDrawings. Select the Publish eDrawing Tool, or select File: Publish eDrawing from the main menu. An eDrawing is created and the eDrawings Publisher is opened, as shown in Figure 2.75.

Note: Once you have installed the eDrawings Publisher to your computer, you can open the program from the Start menu on your computer or an icon on your desktop. Installing eDrawings as a SolidWorks add-in simply provides easy access to the program from the SolidWorks® environment.

By default, the eDrawing views are displayed in the shaded mode. The display mode can easily be changed to show line drawings.

Click on the Shaded icon.

The eDrawings Viewer allows drawings to be animated.

FIGURE 2.75

FIGURE 2.76

Click the Next Tool repeatedly to switch from one view to another, including section and detail views.

When the Section View is active, select the Rotate Tool. Click and drag to rotate the model view, as shown in Figure 2.76.

Select the Home Tool to return to an overview of the drawing.

An important feature of eDrawings is that comments can be added by reviewers. To illustrate, we will add a comment for draft to be added to the outer surface of the 2.75-inch diameter boss.

If the markup tools are not visible at the right side of the screen, select the UI (User Interface) Mode Tool.

This tool allows you to toggle between a simplified user interface (without markup tools) and the complete user interface. To add a comment, we need to have the complete user interface visible.

FIGURE 2.77

FIGURE 2.78

Select the Markup tab from the eDrawing Manager at the left side of the screen, as shown in Figure 2.77. Select the Options button, as shown in Figure 2.78.

FIGURE 2.79

In the box that appears, you can set preferences for your comments. Several reviewers can add comments to the same drawing, and each reviewer can have a unique color.

Enter a name for the reviewer, and select a color and font type/size, as shown in Figure 2.79.

 From the markup tools, select the Cloud with Leader Tool. Click on the edge to be drafted, as shown in Figure 2.80. In the pop-up box that appears, type in the text of the comment, as shown in Figure 2.81.

FIGURE 2.80

Click the check mark to close the box, and then click at the location where the comment is to be placed, as shown in Figure 2.82.

Comments are listed on the left side of the screen, as shown in Figure 2.83, allowing the person making the corrections to the drawing to see who made each comment. In this way, eDrawings allow efficient collaboration among different groups working on a project.

FIGURE 2.82

FIGURE 2.81

FIGURE 2.83

FIGURE 2.84

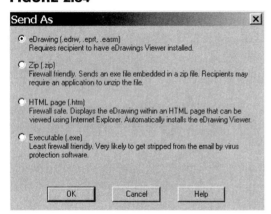

There are several options for saving and sending eDrawings files. Files can be e-mailed directly from the eDrawings Publisher.

Select the Send Tool.

Several options are available, as shown in Figure 2.84. To send the drawing to someone without the eDrawings Viewer, it is necessary to save an executable file, which includes the Viewer. Since the executable file contains the eDrawing Viewer, the file is much larger than the file saved with the .edrw format. In this example, the .exe file is 2100 kB, as compared with the 43 kB file size of the .edrw file. Since many companies have firewalls that prevent .exe files from being received (because of the possibility of viruses), it may be necessary to send the file as a zipped file or as an html. Both of these options result in the eDrawings Viewer being sent along with the eDrawing.

Select cancel to close the Send box. Select File: Save from the main menu, save the eDrawing as an .edrw file, and close the eDrawings Publisher.

Problems

P2.1   Create a SolidWorks drawing of the beam section described in Problem P1.2. Include front and side views in your drawing.

P2.2   Create a SolidWorks drawing of the pulley described in Problem P1.3. Include front and side views in your drawing, as well as a section view. Show a detail view of the keyway region.

P2.3   Create an eDrawing of the pulley from Problem P2.2.

P2.4   Create a SolidWorks drawing of the plastic pipe tee described in Problem P1.5. Include front and side views in your drawing, as well as a section view.

P2.5   Create an executable eDrawing of the pipe tee from Problem P2.4. (*This problem requires the eDrawing add-in.*)

P2.6   Create a SolidWorks drawing of the brake rotor described in Problem P1.6. Include front and side views in your drawing.

P2.7   Create an eDrawing of the brake rotor from Problem P2.6.

Additional Part Modeling Techniques

Introduction

The flange part modeled in Chapter 1 required the use of a number of basic solid modeling construction techniques. In this chapter, some additional construction and dimensioning techniques will be explored. The part that will be modeled in this chapter is a wall-mounted bracket, as shown in Figure 3.1.

FIGURE 3.1

This bracket has some features that will allow us to explore some new modeling techniques:

- It has one plane of symmetry, allowing us to exploit the mirroring capabilities of the program.
- It has a linear hole pattern, which can be readily constructed using the SolidWorks program.
- It has three reinforcing ribs, which can be created with the extruded boss command used earlier or with a rib tool.

In this chapter, a tutorial for modeling this bracket will be presented. Before beginning this tutorial, we will learn how to customize the SolidWorks environment.

In this chapter, you will:

- learn how to customize the SolidWorks environment by modifying the toolbars and the CommandManager,
- learn how to save preferences for options in template files,
- construct centerlines, and use them to locate sketch features through the addition of relations,
- use the Mirror Entities Tool to create symmetric sketches about a centerline,
- perform a new type of extrusion called a Mid-Plane Extrusion,
- add ribs with the Rib Tool,
- use the Mirror Tool to create symmetric features about a symmetry plane, and
- create a linear hole pattern.

3.1 Customizing the SolidWorks Environment

Open a new part file. Click on the Front Plane in the FeatureManager to select it.

We will begin by drawing a circle. However, rather than selecting the Circle Tool from the Sketch group of the CommandManager, as we did in Chapter 1, we will select the Circle Tool from the main menu.

From the main menu, select **Tools: Sketch Entities: Circle,** as shown in Figure 3.2. Drag out a circle, as shown in Figure 3.3 (the size of the circle is not important).

FIGURE 3.2

FIGURE 3.3

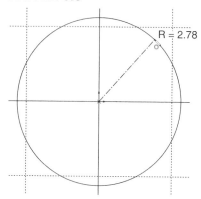

Press the esc key to turn off the Circle Tool. From the main menu, select **Insert: Boss/Base: Extrude,** as shown in Figure 3.4. Click the check mark in the Property Manager to create a solid disk, as shown in Figure 3.5.

FIGURE 3.4

FIGURE 3.5

FIGURE 3.6

The edges of the disk might look rough on the screen. For a smoother appearance, we will adjust the image quality.

Select **Tools: Options** from the main menu. Under the Document Properties tab, select Image Quality. Move the top slider bar toward the right, as shown in Figure 3.6. Click OK to close the dialog box.

FIGURE 3.7

The disk now appears much smoother, as shown in Figure 3.7. The disadvantage of setting the image quality higher is that your computer performance will be slower when refreshing the screen. This performance decrease will be especially noticeable when working with large assemblies or with very complex parts. However, for the parts and assemblies in this book, performance will not be noticeably slower if the image quality is set as high as possible, as long as you are using a relatively fast computer with a good graphics card.

Now we will add a cylindrical boss, using the toolbars to specify sketch and feature commands.

FIGURE 3.8

Right-click on any of the open toolbars (but not the CommandManager). A list of available toolbars is displayed, with open toolbars indicated by a check mark. Click on Features, as shown in Figure 3.8. Right-click on a toolbar again, and click on Sketch.

The Features and Sketch toolbars are now displayed. By default, the Features toolbar will be on the left side of the screen, while the Sketch toolbar will be on the right side.

FIGURE 3.9

FIGURE 3.10

FIGURE 3.11

Select the front face of the part, as shown in Figure 3.9. Click on the Circle Tool in the Sketch toolbar, as shown in Figure 3.10.

Drag out a circle, as shown in Figure 3.11.

FIGURE 3.12

Select the Extruded Boss/Base Tool from the Features toolbar, as shown in Figure 3.12. Enter a distance for the circle to be extruded, and click the check mark to complete the extrusion, which is shown in Figure 3.13.

FIGURE 3.13

We have now activated sketch and feature tools in three different ways:

- from the CommandManager
- from the Main Menu
- from Toolbars

The CommandManager is a new SolidWorks feature, appearing for the first time in the SolidWorks 2004 release. Many experienced users prefer to use the toolbars instead, since they are accustomed to doing so. There is another advantage to using the toolbars: tools can be accessed with a single mouse click. With the CommandManager, it is necessary to select either the Sketch group or the Features group before selecting the desired tool, if the appropriate group is not already displayed. The CommandManager can be turned off and on in the same manner as any toolbar, by right-clicking on a toolbar and clicking on CommandManager on the menu that appears.

In this book, we have chosen to use the CommandManager for sketch and feature tools because the tool names are displayed along with the icons. For new users, this is a helpful feature. We use the Standard, View, and Standard Views toolbars because many of the commands on these toolbars can be used at the same time sketch or feature commands are being used. For example, we might begin an extrusion command from the CommandManager, and realize that a different view mode or orientation might be helpful in setting the extrusion parameters. We can click on any of the view tools from the toolbar without interrupting the extrusion process. We use the main menu to call up commands that are not commonly used. For example, we may want to add an ellipse to a sketch. Rather than adding the Ellipse Tool to the Sketch toolbar, we select Tools: Sketch Entities: Ellipse from the main menu. However, if the types of parts that you model use ellipses regularly, then the Ellipse Tool can be added to the toolbar. We will now learn how to modify toolbars.

FIGURE 3.14

Right-click on any of the toolbars (but not the CommandManager). Near the bottom of the menu that appears, click on Customize, as shown in Figure 3.14.

Move the cursor over the RealView Graphics Tool on the View toolbar, as shown in Figure 3.15. Click and drag the icon into the modeling area, as shown in Figure 3.16.

This will remove the RealView Graphics Tool from the menu. If you have a graphics card that supports RealView, then you might want to add it back later.

FIGURE 3.16

FIGURE 3.15

We will now add a tool to a toolbar. The following step will add the Isometric View Tool to the Standard Views toolbar. If this tool already appears on your computer, you may duplicate the steps shown with any other tool.

In the dialog box, click on the Commands tab and select Standard Views from the list of categories.

A group of icons representing the available tools for this category is displayed. Moving the cursor over each icon displays the tool name and gives a brief description of that tool, as shown in Figure 3.17.

Click on the Isometric Tool and drag it onto the Standard Views toolbar, next to the Trimetric View Tool, as shown in Figure 3.18.

FIGURE 3.17

FIGURE 3.18

FIGURE 3.19

Click OK to close the dialog box. The Isometric View Tool now is included on the Standard Views toolbar, as shown in Figure 3.19.

In addition to demonstrating how to add a tool to a toolbar, the Isometric View Tool was added because you may find that you prefer an isometric view over the trimetric view that was included on the Standard Views toolbar by default. You may have noticed when selecting the Isometric View Tool that there is a third 3-D view option available, the Diametric View.

To understand the differences in the three 3-D views available, consider the cube displayed in each orientation in Figure 3.20. An isometric view displays the part so that edges corresponding to the x, y, and z directions are the same length on the screen (the "iso-" prefix means "equal"). Previous versions of the program used an isometric view as the default 3-D view orientation.

In the SolidWorks 2004 released, two additional 3-D views were added to the Standard Views. A diametric view keeps edges corresponding to two of the three principal directions the same length on the screen, and allows the third edge length to be different. The Diametric View shows the front and right sides of a part more prominently than the top side. A trimetric view is one in which edges of the cube corresponding to the three principal axes are displayed as different lengths. The Trimetric View displays the front

FIGURE 3.20

Isometric   Diametric   Trimetric

FIGURE 3.21

view most prominently. You might find that certain part models are better visualized in one of the three 3-D views. In this book, we use Trimetric View as the default 3-D view.

We will see how to customize the CommandManager. Any toolbar can be shown on the CommandManager.

Right-click on the CommandManager, and select Customize CommandManager from the menu that appears, as shown in Figure 3.21.

FIGURE 3.22

A list of toolbars is displayed, with the toolbars currently on the CommandManager indicated with check marks, as shown in Figure 3.22. Click the box for the Standard Views toolbar, and then click anywhere in the modeling area to close the menu.

The Standard Views Tools are now available from the Command-Manager, as shown in Figure 3.23. Note that the tools are the same as those on the toolbar. When tools are added or removed from the toolbar, then those changes are reflected in the CommandManager.

Right-click on the CommandManager, and select Customize CommandManager. Click on the Standard Views to remove the check mark, and click in the modeling area to close the menu.

As noted earlier, it is easier to work with tools directly from the menus, and we will continue to do so with the Standard Views Tools. In a later chapter, when making a sheet metal part, we will add the Sheet Metal Tools to the CommandManager.

Note that when you make changes to a toolbar, you do not need to save the changes. The modifications will be automatically saved.

Document options are not automatically saved. For example, if the default system of units is millimeters/grams/seconds and you usually work in inches/pounds/seconds, then you must change the options for every new part. However, you can change the default options by modifying the template in which they are saved. The following steps demonstrate the modification of the default part template. If you use a shared computer, then you may prefer to change the options every time and leave the defaults as set.

FIGURE 3.23

Optional steps for modifying the part template:
Open a new part file. From the main menu, select Tools: Options, as shown in Figure 3.24.

FIGURE 3.24

Under the Document Properties tab, set the following options as desired (recommended values in parentheses):

- Grid/Snap (show grid)
- Units (inches/pounds/seconds), number of decimal places = 2 or 3
- Colors (a light color for the Shading color)
- Image Quality (move slider bar to High)

Click **OK** to close the Options box.

From the main menu, select File: Save As. In the dialog box, use the pull-down menu to select Part Templates as the file type, as shown in Figure 3.25. Click on the file named "Part.prtdot" to select it, and click Save. When asked if you want to replace the default template, click Yes. Close the part file.

FIGURE 3.25

When you open a new part file, the settings just saved will be applied. You may also save templates with other names. For example, you can have a "Part mm" template for use when modeling a part in millimeters and a "Part in" template for modeling a part in inches. Templates are especially useful with drawings. You can save templates for different unit systems, drawing sizes, and title blocks.

3.2 Part Modeling Tutorial: Bracket

Open a new part file. Be sure that dimensions are set to inches (with two decimal places displayed) and the Display Grid option is on.

FIGURE 3.26

Click on the Front Plane in the FeatureManager to select it, as shown in Figure 3.26.

Click on the Sketch icon so that the Sketch group of tools is displayed in the Command-Manager. Click the Rectangle Tool.

Sketch

Rectangle

FIGURE 3.27

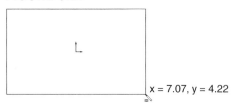

x = 7.07, y = 4.22

FIGURE 3.28

FIGURE 3.29

FIGURE 3.31

FIGURE 3.32

Now left-click anywhere in the drawing area and while holding down the left mouse key, drag out a rectangle, as shown in Figure 3.27.

Don't worry about the locations of the corner points; we will add dimensions that will size and locate the rectangle.

Click on the Smart Dimension Tool, and dimension the horizontal side of the rectangle to 6 inches and the vertical side to 4 inches, as shown in Figure 3.28.

Note that the sides of the rectangle are shown blue, and the status bar at the lower-right corner of the screen shows that the sketch is "Under Defined." Although we have fully defined the size of the rectangle, we have not specified its position. We can do this by adding dimensions from the origin to two of the sides, or by using a centerline for construction. We will use the centerline approach.

Click on the Centerline Tool. With the left mouse button depressed, drag a diagonal line, snapping from corner to corner of the rectangle.

The dimensioned sketch with the centerline added should appear as shown in Figure 3.29.

A relation will now be added to locate the rectangle so that it is centered at the origin.

Click the esc key to turn off the Centerline Tool. Hold down the ctrl key, and use the left mouse button to select both the centerline and the origin.

Recall that the ctrl key is used to make multiple selections.

In the Property Manager, select Midpoint to add a relation between the centerline and the origin, as shown in Figure 3.30.

The rectangle is now both dimensioned and located, so the sketch is fully defined.

Click on the Features icon and then the Extruded Boss/Base Tool. In the PropertyManager, enter 0.25 inches as the thickness, and click on the check mark, as shown in Figure 3.31.

A solid object has been created, as shown in Figure 3.32.

FIGURE 3.30

The horizontal boss of the bracket will now be constructed. The feature will be sketched in a plane 1.25 inches below the top surface of the base part. To accomplish this, we will first need to create the reference plane to be used for sketching.

FIGURE 3.33

Select the top surface of the base part (see Figure 3.33).

Be sure that the face is selected (a flag symbol will appear next to the cursor).

From the Features group of the CommandManager, select Reference Geometry: Plane, as shown in Figure 3.34. In the PropertyManager, click the check box to reverse the direction, and set the distance to 1.25 inches (see Figure 3.35). Click the check mark to create the plane, which will be named Plane1.

FIGURE 3.34

FIGURE 3.35

The boss feature will now be sketched in this new plane. The symmetry of the boss about the Right Plane will be exploited in the creation of the sketch.

FIGURE 3.36

Select the Top View. Use the Pan Tool to place the part toward the top of the screen, as shown in Figure 3.36. Press the esc key to turn off the Pan Tool.

We will use the Mirror Tool to assist in construction of the symmetric sketch. This will require that we draw a centerline first.

FIGURE 3.37

If Plane1 is not selected (highlighted in green), then click on Plane1 in the Feature-Manager. Select the Centerline Tool from the Sketch group of the CommandManager, and drag a line from the origin downward. As you drag the line downward, a "V" by the cursor shows that the line will be vertical (see Figure 3.37).

Sketch

Centerline

Now click the Mirror Entities Tool.

Drawing features added now will be mirrored about the centerline.

Mirror Entities

FIGURE 3.38

Select the Line Tool (solid line, not the centerline), and move the cursor near the edge of the previously created solid.

Line

A lightbulb symbol will appear next to the cursor to show that you are snapping to the edge, as shown in Figure 3.38.

FIGURE 3.39

While holding down the left mouse key, drag a line downward at an angle (Figure 3.39).

When you release the mouse key, you will see that the line you were drawing and another line that is a mirror image are created.

Select the Tangent Arc Tool. Move the cursor to the endpoint of the line, so that the square appears next to the cursor, indicating that the arc will begin at the endpoint of the line (see Figure 3.40).

FIGURE 3.40

FIGURE 3.41

A = 103.45°
R = 0.96

Tangent Arc

Hold down the left mouse button and drag out an arc, with the endpoint on the centerline, as shown in Figure 3.41. Click on the Mirror Tool icon to turn the mirroring off.

The arc that you drew will remain tangent to the first line, no matter how the dimensions of the sketch are changed. Your sketch should now appear as shown in Figure 3.42. Of course, this is not exactly what we want: the two arcs should be tangent to each other. Therefore, we will add a relation to the two arcs.

FIGURE 3.42

FIGURE 3.43

Select the two arcs, and click Concentric in the Property-Manager, as shown in Figure 3.43. Click the check mark to add the relation.

The two arcs are now tangent to each other, as shown in Figure 3.44. Note that we could have specified a tangent relation instead of the concentric relation, but there are two possible solutions to the tangent relation. To ensure that the geometry is the way we want it, the concentric relation provides only one solution.

Chapter 3 Additional Part Modeling Techniques 75

FIGURE 3.44

FIGURE 3.45

FIGURE 3.46

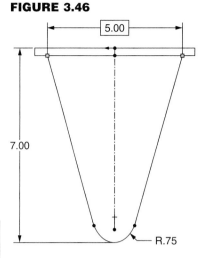

Select the Line Tool, and finish the sketch with a horizontal line connecting the two free ends of the sketch, as shown in Figure 3.45. Select the Smart Dimension Tool, and dimension the sketch as shown in Figure 3.46.

Note that the 7-inch dimension is to the *back* of the bracket.

Select the Extruded Boss/Base Tool from the Features Group of the CommandManager. Click the Reverse Direction icon in the PropertyManager, as shown in Figure 3.47, and extrude the part downward 0.25 inches. Change to the Trimetric View to preview the extrusion, and click the check mark to complete the operation.

FIGURE 3.47

The bracket should appear as shown in Figure 3.48.

Now we will add a raised cylindrical feature for reinforcement near what will become a mounting hole.

FIGURE 3.48

Select the top surface of the horizontal extruded boss, as shown in Figure 3.49. Select the Top View.

We want to locate the raised area concentric with the rounded tip of the boss. In order to establish this relation, we will first need to "wake up" the arc that makes up this rounded tip, allowing the center point to be used as a snap point in our sketch.

FIGURE 3.49

Select the Circle Tool from the Sketch group of the Command-Manager. Move the cursor into the sketching window, and hold the pointer (without clicking) over the arc (see Figure 3.50, which is shown in wireframe mode for clarity).

FIGURE 3.50

The center point of the arc now appears, and it can be used as a snap point for the center of the circle.

FIGURE 3.51

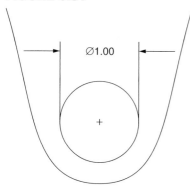

Drag out a circle, starting at the center mark. Select the Smart Dimension Tool, and dimension the circle to be 1 inch in diameter (see Figure 3.51).

Select the Extruded Boss/Base Tool from the Features group of the CommandManager. Extrude the sketch upward for a distance of 0.050 inches, as shown in Figure 3.52.

A reinforcing rib will now be added to the part. This will be created using a new type of extrusion, called a Mid-Plane Extrusion.

FIGURE 3.52

Select the Right Plane from the FeatureManager, and select the Right View. Select the Line Tool from the Sketch group of the CommandManager, and sketch in the line shown in Figure 3.53. The endpoints of your line should snap to the edges of the solids, but not to any specific point on the edges. Recall that a lightbulb icon next to the cursor indicates that you are snapping to the edge.

FIGURE 3.53

FIGURE 3.54

Add one horizontal and one vertical line coincident with the edges of the solid (snap to the endpoints of the first line and the intersection point of the edges), completing the profile of the rib, as shown in Figure 3.54.

Select the Smart Dimension Tool, and dimension the sketch as shown in Figure 3.55.

FIGURE 3.55

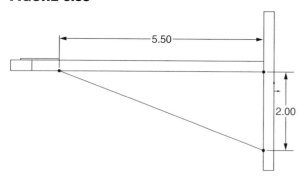

Select the Extruded Boss/Base Tool from the Features group of the CommandManager. In the PropertyManager, select Mid Plane as the type from the pull-down menu (see Figure 3.56), and enter a thickness of 0.1875 inches (see Figure 3.57). Click the check mark to complete the extrusion.

FIGURE 3.56

Unlike the one-sided blind and "up to next" extrusions used previously, the mid-plane type extrudes material in *both* directions, from the sketch plane outward. For a Mid-Plane Extrusion, the thickness entered is the *total* thickness of material after the extrusion. A front view of the part shows that the rib is symmetric about the Right Plane (see Figure 3.58).

FIGURE 3.57

FIGURE 3.58

FIGURE 3.59

FIGURE 3.60

FIGURE 3.62

FIGURE 3.61

Sketch   Line

Smart
Dimension

Features   Rib

The next rib will be added in a slightly different manner, using the Rib Tool. But first, we must create a new plane for the rib. So far, we have used the Insert: Reference Geometry: Plane command from the main menu to create a plane. Here, we will use a drag-and-drop technique that is similar to that used in other Windows applications.

Select the Trimetric View. Click on the Right Plane in the FeatureManager to select it. Hold down the ctrl key, and move the cursor over one of the lines defining the plane until the move arrows appear, as shown in Figure 3.59. Click and drag the mouse to the right. When the mouse button is released, a new plane is created, as shown in Figure 3.60. In the PropertyManager, set the offset distance to 1 inch, as shown in Figure 3.61. Click OK to complete the definition of the new plane, which will be labeled Plane2.

Select the Line Tool from the Sketch group of the CommandManager. Draw a diagonal line as shown in Figure 3.62, snapping the endpoints to the edges of the solids (but not to any specific points).

Select the Smart Dimension Tool and add the dimensions shown in Figure 3.63.

For the first rib, we added two lines to close the sketch and used an Extruded Boss command to complete the rib. A rib can be created from an open sketch using the Rib Tool.

Select the Rib Tool from the Features group of the CommandManager. In the FeatureManager, set the thickness to 0.1875 inches, as shown in Figure 3.64.

FIGURE 3.63

FIGURE 3.64

FIGURE 3.65

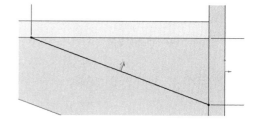

The thickness can be set to be offset from the plane in either direction, or so that the rib is symmetric about the plane (similar to the Mid-Plane Extrusion used for the first rib). By default, the middle of the three buttons, designating a rib symmetric about the plane, is selected. For the extrusion direction, the rib can be defined as parallel to the plane of the sketch (default) or normal to the plane. Neither of these settings needs to be changed for our rib.

FIGURE 3.66

Check the "Flip material side" box, as shown in Figure 3.64, so that the arrow defining the rib direction points toward the part, as shown in Figure 3.65. Click the check mark to complete the rib, which is shown in Figure 3.66.

Earlier, we used the Mirror Entities Tool to create symmetric elements within a sketch. A similar tool, the Mirror Tool, is used to create symmetric features in a part. We will use this tool to create the final rib.

Select the Mirror Tool from the Features group of the CommandManager. In the PropertyManager, the Mirror Face/Plane box is active (highlighted in pink), as shown in Figure 3.67. Click on the Right Plane in the FeatureManager, which has "flown out" to the right of the PropertyManager. The rib just created should be listed as the feature to mirror. If it is not, click that box to make it active and select the rib from the FeatureManager.

FIGURE 3.67

A preview of the final rib is displayed, as shown in Figure 3.68.

Click the check mark to complete the rib.

The three ribs are shown in Figure 3.69. The holes will now be added to the part.

FIGURE 3.68

FIGURE 3.69

Select the top surface of the raised circle, as shown in Figure 3.70. Select the Circle Tool from the Sketch group of the CommandManager. Select the top view.

FIGURE 3.70

As before, it is necessary to "wake up" a center mark at which the hole will be centered.

FIGURE 3.71

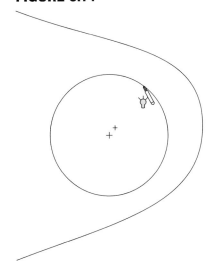

Move the cursor to the edge of the rounded end of the part, or the circular edge of the reinforced area, without clicking a mouse button. This will cause the center mark to be displayed, as shown in Figure 3.71.

FIGURE 3.72

Click and drag out a circle from the center mark. Select the Smart Dimension Tool, and dimension the diameter of the circle as 0.50 inches, as shown in Figure 3.72.

Ø.50

Select the Extruded Cut Tool from the Features group of the CommandManager, and select Through All as the type. Click the check mark to complete the hole, which is shown in Figure 3.73.

Features

Extruded Cut

FIGURE 3.73

The four-hole bolt hole pattern used to mount the flange to the wall will now be added. This will be done by creating a single hole, and then defining a *pattern* based on this hole. In Chapter 1, a *circular pattern* was used to create a bolt hole pattern in the flange. In this exercise, a new type of pattern known as a *linear pattern* will be introduced.

The first step will be the creation of a single bolt hole.

FIGURE 3.74

FIGURE 3.75

⌀.38

FIGURE 3.76

.50

⌀.38

.50

Sketch   Circle

Smart
Dimension

Select the large vertical face of the base part as shown in Figure 3.74. Select the Circle Tool from the Sketch group of the CommandManager, and sketch a circle near the upper-left corner of the face. Select the Smart Dimension Tool and dimension the hole diameter as 0.375 inches, as shown in Figure 3.75.

Note that the diameter dimension is displayed to two decimal places, since that is the option that we set for this part. However, the dimension is stored to the same accuracy we entered.

Add linear dimensions between the circle and the edges, as shown in Figure 3.76.

FIGURE 3.77

Select the Extruded Cut Tool from the Features group of the CommandManager, and select Through All as the type. Click the check mark to complete the hole, which is shown in Figure 3.77.

Features   Extruded Cut

With this single hole defined, a linear hole pattern can now be created.

With the new hole selected, select Insert: Pattern Mirror: Linear Pattern from the main menu, as shown in Figure 3.78.

The feature Cut-Extrude2 will be shown in the "Features to Pattern" box,

FIGURE 3.78

FIGURE 3.79

.50

⌀.38

.50

Extrude1

indicating that this is the base feature for the linear pattern. The "Direction 1" box in the PropertyManager will be highlighted in pink; this allows the first linear direction of the pattern to be established.

Click on the top horizontal edge of the rectangular base to establish the first direction, as shown in Figure 3.79.

In the PropertyManager, set the Direction 1 distance to 5 inches and the number of instances to 2, as shown in Figure 3.80. The Direction 2 box in the PropertyManager will be highlighted pink.

FIGURE 3.80

FIGURE 3.81

FIGURE 3.82

A preview of a second patterned hole will appear, as shown in Figure 3.81.

Click on the left vertical edge of the rectangular base to establish the second direction of the pattern, as shown in Figure 3.82. In the PropertyManager, set the Direction 2 distance to be 3 inches and the number of instances to be 2, as shown in Figure 3.83.

FIGURE 3.83

If the preview of the patterned holes shows that the pattern is created in the wrong direction, click on the arrow at the end of the edge defining the pattern direction, as shown in Figure 3.84.

When the preview of the pattern appears correct, as shown in Figure 3.85, click the check mark to complete the pattern.

FIGURE 3.84

The completed part is shown in Figure 3.86.

Save the part file.

FIGURE 3.85

FIGURE 3.86

DESIGN INTENT

In this chapter, we used mirroring techniques for sketch entities and features to define symmetric elements in a part. The four mounting holes in the bracket, which were created with a linear pattern, could have been defined with mirror commands. The advantage of doing so would be that if we change the dimensions of the vertical portion of the bracket (6 inches by 4 inches), the hole positions would be updated to the correct positions. The same associativity between the hole locations and the size of the bracket can be added if a linear pattern is used, but will require the addition of equations, which will be explained in Chapter 4.

Using symmetric elements in a part is good modeling practice. If a part contains planes of symmetry, plan your model to take advantage of those planes. In the bracket model, we centered the initial sketch about the origin. This placed the part so that the Right Plane coincided with the symmetry plane of the entire part, and the Top Plane coincided with an additional symmetry plane of the vertical portion of the part.

Problems

P3.1   Create a solid model of a structural tee section by creating the sketch shown in Figure P3.1a on the Front Plane and mirroring it about the centerline. Use relations as appropriate to fully define the sketch. Extrude it using a Mid-Plane Extrusion to a total thickness of 36 inches to create the model shown in Figure P3.1b.

FIGURE P3.1a

FIGURE P3.1b

P3.2   The cross-section of a W14×370 Wide-Flange Beam is shown in Figure P3.2a. Create a model of a 36-inch long segment of a W14×370 beam by creating the sketch shown in Figure P3.2b and mirroring it about the two centerlines. (Note: This must be done in two steps. Select the items to be mirrored and one of the centerlines, and then select the Mirror Entities Tool. Then select all entities to be mirrored and the other centerline and select the Mirror Entity Tool. Only one centerline can be selected for each mirror operation.) Extrude the sketch to yield the beam section shown in Figure P3.2c. Set the material type to Plain Carbon Steel and find the weight of the beam segment.

(Answer: 1101 lb)

FIGURE P3.2a

FIGURE P3.2b

FIGURE P3.2c

P3.3   Modify the wide-flange beam segment of P3.2 by adding six 1 inch-diameter holes to each end of the beam, as shown in Figure P3.3a. The locations of the holes are shown in Figure P3.3b. Create a single hole, and use a linear pattern to place the other five holes in one end. Then use a mirror command to place the holes in the other end.

FIGURE P3.3b

FIGURE P3.3a

P3.4 Create a part model of the plastic bracket shown in Figure P3.4a and detailed in Figure P3.4b. Use symmetry in your model so that if you change the width of the part from 2 to 3 inches, the rib and hole placements remain symmetric, as shown in Figure P3.4c.

FIGURE P3.4b

FIGURE P3.4a

FIGURE P3.4c

P3.5 Create a model of the perforated board shown in Figure P3.5a by the following procedure:

a. Begin with a 4 inch by 2 inch sketch, centered at the origin using a centerline and a relation (Figure P3.5b).

FIGURE P3.5a

FIGURE P3.5b

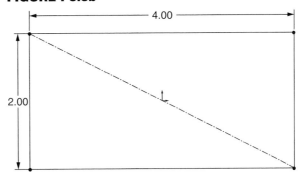

b. Extrude the sketch 0.1 inch.

c. Create and locate a square thru-hole, as shown in Figure P3.5c. (Note: Add a point to the intersection of the diagonal centerlines so that you can dimension to the center of the rectangle.)

d. Use a linear pattern to create an evenly spaced 40×20 grid of holes.

P3.6 Create a solid model of a hacksaw blade (as shown in Figure P3.6a, with a close-up view of the teeth shown in Figure P3.6b), using a linear patterned cut to create the saw teeth. Follow the procedure outlined below.

a. Begin by creating a sawblade "blank," using the dimensions shown in Figure P3.6c and extruding the shape to a 0.02 inch depth.

b. Sketch a single tooth profile, and extrude a cut through the blank (see Figure P3.6d).

c. Create a linear pattern to copy the tooth profile the length of the sawblade.

FIGURE P3.5c

FIGURE P3.6b

FIGURE P3.6a

FIGURE P3.6c

FIGURE P3.6d

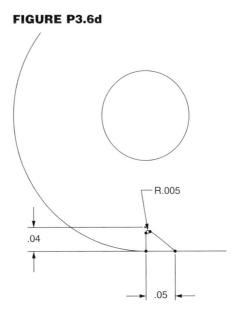

R.005

.04

.05

P3.7   Create a solid model of the muffin pan shown in Figure P3.7a and detailed in Figure P3.7b. The entire pan is 0.050 inches thick. Each of the 12 muffin wells is 3 inches in diameter at the top, is 1.5 inches deep, and has tapered sides, as shown in Figure P3.7c. The entire top surface and the insides of the muffin wells are to be coated with a nonstick material. Find the surface area that will be coated. [Hint: The Mass Properties Tool gives you the surface area of the entire part, but to get the surface area for a single surface or a specific group of surfaces, select the surface(s), using the ctrl key to select multiple items, and select Tools: Measure from the main menu.]

FIGURE P3.7a

FIGURE P3.7b

10.50

A   A

Ø3.00 typ.

14.00

3.50 typ.

R1.00 typ.

3.50 typ.

FIGURE P3.7c

(3.00)

.05

Taper sides
10 deg

1.50

.05

Section A-A

Use of Parametric Modeling Techniques

Introduction

One of the tremendous advantages of solid modeling software is the ability of the software to represent geometric relationships in a *parametric* form. In a parametric model, the description of the model contains both a geometric description (shape, interconnectivity, etc.) and specific numerical parameters (dimensions, number of instances, etc.). Consider, for example, a simple parameterized solid model of a cylinder; the definition of the model contains both a geometric description defining the primitive shape (cylinder), and two numeric parameters (radius and length) to fully define the model. A whole "family" of cylindrical parts of different radii and lengths could be modeled with a single parameterized model, with only a numeric table of values used to differentiate the various parts in the family.

The Linear Pattern Tool used to define the hole pattern in Chapter 3 is another good example of a parameterized model; the geometric information used to describe a hole pattern includes:

- the shape of the hole (circular)
- the type of extruded cut ("through all")

The numeric parameters that must also be defined include:

- the vectors (axes) that define the two directions of the pattern
- the two "repeat" dimensions along the two directions
- the number of instances of the holes in each direction

Changes can be readily made to the parameters to update the model, without requiring any changes to the geometric information.

This concept of using numeric parameters to drive a solid model can be exploited in mechanical design. One powerful tool that can be employed is the development of mathematical relationships (equations) that relate the values of two or more parameters in a model. In our hole pattern, for example, we might want to restrict our model so that the number of instances of holes in one direction is always the same as the number of instances of holes in the other direction. This is accomplished by relating the two parameters together with an equation, for example:

(# of instances along Direction 2) = (# of instances along Direction 1)

Once this parameterized relationship is established, any change made to the *independent parameter* (the number of instances of holes along Direction 1) would automatically update the value of the *dependent parameter* (the number of instances of holes along Direction 2), and the model would be updated accordingly. This allows the engineer to embed design intelligence and "rules-of-thumb" directly into solid models.

In this chapter, an example of a mechanical part that contains embedded equations to drive specific parameters will be presented. In addition, the tutorial will introduce some new modeling operations that have not yet been introduced in the preceding chapters.

In this chapter, you will:

- learn how to view a section view of a model,
- use the Convert Entities Tool to convert the silhouette of a solid part into a sketch line,
- use the Draft Tool to add draft to an existing solid part,
- perform a revolved cut operation,
- develop equations to embed parameterized relationships into a solid model,
- use a design table to create a family of similar parts, and
- create a multiple-configuration part drawing.

4.1 Modeling Tutorial: Molded Flange

In this exercise, we'll create the flange shown here in Figure 4.1. Since this is to be a molded part, we'll add draft to the appropriate surfaces.

FIGURE 4.1

Begin by starting a new part, and selecting the Front Plane from the FeatureManager. Select the Circle Tool from the Sketch group of the Command-Manager. Drag out a circle centered at the origin. Select the Smart Dimension Tool and add a 6-inch diameter dimension, as shown in Figure 4.2.

Select the Extruded Boss/Base Tool from the Features group of the CommandManager. Extrude the sketch outward to a thickness of 0.25 inches.

The extruded solid is shown in Figure 4.3.

Since the next feature will be drafted, we will define its dimensions in a new plane and extrude its sketch back toward the rest of the part. Therefore, we need to create a new construction plane offset from the Front Plane.

Begin by selecting the Front Plane from the Feature-Manager.

The selected plane will be highlighted in green.

While holding down the ctrl key, click and drag the plane outward. Make sure that the move arrows, as shown in Figure 4.4, appear before clicking. Release the ctrl key, and the new plane is previewed, as shown in Figure 4.5.

In the PropertyManager, set the offset distance to 1.5 inches, as shown in Figure 4.6, and click the check mark to complete the operation.

FIGURE 4.2

Ø6.00

FIGURE 4.3

FIGURE 4.4

Front Plane

Front Plane

FIGURE 4.5

FIGURE 4.7

The new plane will be labeled Plane1 in the Feature-Manager. A cylindrical boss will now be extruded with a draft from this new plane back toward the base feature.

With the new plane (Plane1) selected, select the Circle Tool from the Sketch group of the CommandManager. Drag out a circle from the origin. Select the Smart Dimension Tool and dimension the circle's diameter as 2.2 inches, as shown in Figure 4.7.

Select the Extruded Boss/Base Tool from the Features group of the CommandManager, and set the direction such that the boss is extruded toward the base. Change the type of extrusion to Up to Next. (Note: If Up to Next is not available, then the extrusion direction is incorrect.) Turn the draft on. Set the draft angle to 3 degrees and check the "Draft outward" box, as shown in Figure 4.8. Click on the check mark to complete the extrusion.

FIGURE 4.6

FIGURE 4.8

Draft

FIGURE 4.9

The result of the drafted extrusion is shown in Figure 4.9.

Select View: Planes from the main menu. Then select View: Origins from the main menu.

These steps will toggle the display of planes and origins from on (check mark shown in menu) to off (no check mark). The part is shown in Figure 4.10 without Plane1 and the origin visible.

FIGURE 4.10

Select the face shown in Figure 4.11. Select the Circle Tool from the Sketch group of the CommandManager. Drag out a circle centered at the origin. Select the Smart Dimension Tool. Dimension the circle's diameter at 1.8 inches, as shown in Figure 4.12.

FIGURE 4.11

FIGURE 4.12

FIGURE 4.13

Features

Extruded
Cut

Select the Extruded Cut Tool from the Features group of the CommandManager. Select Through All as the type, turn the draft on, and set the draft angle to 2 degrees. Make sure that the "Draft outward" box is checked, as shown in Figure 4.13. Click the check mark to apply the cut, which is shown in Figure 4.14.

FIGURE 4.14

To see the effect of the draft on the hole, viewing the model with a section view is helpful.

Select the Section View Tool from the View toolbar. In the PropertyManager, select the middle button representing the Top Plane, as shown in Figure 4.15.

FIGURE 4.15

FIGURE 4.16

The section view of the model is previewed, as shown in Figure 4.16.

Reverse the direction of the section by clicking on the icon next to the plane selection box, as shown in Figure 4.17.

The new preview of the section view is shown in Figure 4.18.

FIGURE 4.18

Click the check mark to apply the settings for the section view. Switch to the Bottom View.

The cross-section of the part, as shown in Figure 4.19, clearly shows the drafted center hole.
The section view will be displayed until turned off with the Section View Tool.

FIGURE 4.17

Click the Section View Tool to return to the display of the entire model.

FIGURE 4.19

Before adding the ribs and holes, we will fillet several of the sharp edges of the part.

Start by selecting the Fillet Tool.

Fillet

Since two of the edges are to have the same radius, we can add these radii at the same time.

FIGURE 4.20

Set the radius to 0.125 inches, and select the two edges shown in Figure 4.20.

Be sure to select the edge and not the face. If a face is selected, then all of the edges of that face will be filleted.

FIGURE 4.21

Using a similar procedure, fillet the edge shown in Figure 4.21 with a fillet radius of 0.25 in.

The filleted part is shown in Figure 4.22.

A number of stiffening ribs will now be added to the part. This will be done by creating a single rib and duplicating it a number of times using a circular pattern. The first rib in the pattern will now be created.

FIGURE 4.22

Select the Top Plane from the FeatureManager. Select the Sketch Tool from the Sketch group of the CommandManager to open a sketch. Choose the Bottom View.

Sketch   Sketch

Choose the Wireframe Tool. Zoom in on the right side of the part, as shown in Figure 4.23.

FIGURE 4.23

Since the rib will blend into the fillet radii, we need to select the intersections of the radii with the plane. While the edge of the solid part appears to be a line when viewed from this perspective, there is no physical sketch line associated with it; however, we can construct a line coincident with this projected solid edge by using the Convert Entities Tool.

FIGURE 4.24

Fillet1

Move the cursor to the intersection of the top fillet and the plane. When you have positioned the cursor at the appropriate location, the silhouette symbol will appear (Figure 4.24). Click to select this silhouette.

With the silhouette selected, click the Convert Entities Tool.

Convert Entities

This will add to your sketch an arc that is coincident with the silhouette of the fillet.

FIGURE 4.25

FIGURE 4.26

Do the same for the other fillet, so that there are now two arcs in the active sketch, as shown in Figure 4.25.

Select the Line Tool. Move the cursor over the top arc so that a lightbulb appears, as shown in Figure 4.26, indicating that the endpoint of the line will snap to the arc.

FIGURE 4.27

Click and drag a line to the other arc, again making sure that the endpoint snaps to the arc, as indicated by a lightbulb, as shown in Figure 4.27. Press esc to turn off the Line Tool.

We want the rib to blend tangent to these fillets. We will ensure this by adding tangent relations between the constructed arcs and the line.

With the ctrl key depressed, select one of the constructed arcs and the line. Click the Tangent box to establish a tangent relation (Figure 4.28), and click the check mark to apply the relation.

FIGURE 4.28

FIGURE 4.29

Repeat this operation with the other arc and the line.

Now that the tangent relations have been established, part of the constructed arcs must be trimmed away.

Zoom in on the intersection of one of the arcs and the line. Select the Trim Entities Tool. Click to trim the portion of the arc to be removed, as shown in Figure 4.29.

FIGURE 4.30

The result of the trimming operation is shown in Figure 4.30.

Repeat this process on the other arc/line intersection.

Add the two lines indicated in Figure 4.31 to close the sketch contour. (The two lines do not have to be vertical and horizontal, but should both lie within the edges of the part, so that there will be no gaps when the sketch is extruded into a solid rib.)

FIGURE 4.31

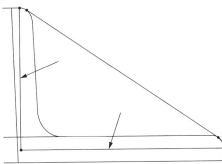

Select the Extruded Boss/Base Tool from the Features group of the CommandManager. Extrude the rib with a Mid-Plane Extrusion with a thickness of 0.125 inches.

Features   Extruded Boss/Base

The result is shown in Figure 4.32.

Since this is a molded part, we will need to add a draft to the rib. However, unlike the base feature and the first extruded boss, the necessary draft on the rib is not in the same direction as the extrusion; therefore, the draft on the rib will be added as a secondary operation.

FIGURE 4.32

Select the Draft Tool.   Draft

The Draft PropertyManager will appear. First we must specify a neutral plane, from where the draft originates. The Neutral Plane box will be highlighted pink.

Select the top of the flange as the neutral plane. Be sure that you select the face, as indicated by the flag symbol.

FIGURE 4.33

Direction of Draft

The arrow points in the direction for which the thickness *decreases* (see Figure 4.33). Therefore, this draft is in the correct direction.

In the Draft PropertyManager, click on the box for the Faces to Draft.

The Faces to Draft box will be highlighted in pink.

Select one side face of the rib.

Use the Rotate View Tool to rotate the view and select the face on the other side of the rib also.

FIGURE 4.34

Set the draft angle at 2 degrees, as shown in Figure 4.34, and click the check mark.

FIGURE 4.35

The effect of the draft can be seen clearly from the front view, as shown in Figure 4.35.

Select the Circular Pattern Tool. Select View: Temporary Axes from the main menu.

Circular Pattern

Both the rib *and* the draft must be copied in the circular pattern.

FIGURE 4.36

FIGURE 4.37

Click on the center axis to define it as the axis of rotation, as shown in Figure 4.36.

Click in the "Features to Pattern" box to activate it, and select both the rib (Extrude3) and the draft from the FeatureManager, as shown in Figure 4.37.

(It may be necessary to click on the plus sign next to the part name to expand the fly-out FeatureManager.)

FIGURE 4.38

Set the number of instances to 6, and click on the check mark to complete the pattern.

FIGURE 4.39

The resulting part is shown in Figure 4.38.

While the geometry of the flange appears to be satisfactory, there is an area where the faces do not blend smoothly. This mismatch occurs at the intersection between the rib and the fillet, as shown in Figure 4.39.

Note that we have a flat part (the rib) mating with a curved part (the fillet). Although they match perfectly at the mid-plane of the rib, as the rib is extruded outward its surface is slightly higher than the fillet's. Although the mismatch is small, this type of mismatch can result in errors when exporting the geometry to a rapid prototyping system, tool path program, or finite element program.

One way to remove the mismatch is to cut away the high surfaces of the rib. Think of this operation as sweeping a cutting tool around the part. We will want our tool to match the contour of the rib mid-plane. This is most easily done by copying the sketch that was used to create the rib.

FIGURE 4.40

In the FeatureManger, click on the plus sign beside the rib (Extrude3) to display the sketch used to create the rib. Select the sketch by name (Figure 4.40). From the main menu, select Edit: Copy.

FIGURE 4.41

FIGURE 4.42

Select the Top Plane, and select Edit: Paste from the main menu.

A new sketch that is identical to the one used to create the rib is placed in the Top Plane.

Right-click on the new sketch in the FeatureManager, and select Edit Sketch (Figure 4.41).

FIGURE 4.43

Delete the horizontal and vertical lines, and add the lines shown to complete the "cutting tool" geometry as shown in Figure 4.42, including a vertical centerline from the origin.

Select the Revolved Cut Tool.

Click the check mark to accept the default setting of 360 degrees and complete the cut, as shown in Figure 4.43.

DESIGN INTENT

When we copied the sketch used to create the rib, the new sketch was underdefined. This is because entities in the original sketch were linked to the silhouettes of the flange edges, but the copied sketch entities were not. The new sketch could not be re-created in the same way as the original sketch, because the silhouette edges were changed when the rib was extruded. One method of using a fully defined sketch for the second sketch would be to create the sketch in the Right Plane rather than the Top Plane. Since there is no rib at that location, the sil-

houette edges could be used to define the sketch entities. However, if the number of ribs is changed later to four or eight, then an error would result because the silhouette edges would not be present.

While ideally we would like to be able to change any parameters of a model and have it rebuild correctly, sometimes that result is difficult to achieve. The intent when creating this part is to allow the number of ribs and holes to be changed. Our method of creating the part allows this intent to be realized.

The result is a smooth transition in the problem areas, as shown in Figure 4.44. Note that the top of the rib is no longer flat, but rather has a slight curvature now.

The holes will now be added.

Select the face shown in Figure 4.45.

FIGURE 4.44

FIGURE 4.45

FIGURE 4.46

Sketch   Circle   Smart Dimension

Select the Circle Tool from the Sketch group of the CommandManager.
Draw a circle centered at the origin. Check the "For construction" box in the PropertyManager. Select the Smart Dimension Tool, and dimension the circle diameter as 4.5 inches, as shown in Figure 4.46.

FIGURE 4.47

Select the Centerline Tool and draw two centerlines, both originating from the origin. One of the centerlines should be horizontal and the other diagonal, as shown in Figure 4.47. Select the Smart Dimension Tool and add a 30-degree angular dimension between the centerlines.

Centerline

Smart Dimension

FIGURE 4.48

Select the Circle Tool. Move the cursor to the intersection of the construction circle and the diagonal centerline, so that the intersection icon appears, as shown in Figure 4.48.

Circle

Drag out a circle. Select the Smart Dimension Tool and add a 0.375-inch diameter dimension to the circle, as shown in Figure 4.49.

Smart Dimension

Select the Extruded Cut Tool from the Features group of the CommandManager. Set the type as Through All, and click the check mark to create the first hole.

Features

Extruded Cut

FIGURE 4.49

Select the Circular Pattern Tool. Specify a six-hole pattern.

Circular Pattern

The finished part is shown in Figure 4.50.

In the next section, we will add equations to link the rib and hole patterns. This will be easier to do if we rename these features.

FIGURE 4.50

In the FeatureManager, rename the extrusion defining the first rib "Rib," the first circular pattern "Rib Pattern," the first bolt hole "Bolt Hole," and the second circular pattern "Hole Pattern," as shown in Figure 4.51. Save the part.

FIGURE 4.51

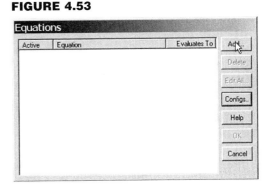

4.2 Creation of Parametric Equations

The use of parametric equations embedded in models to control dimensions is a powerful engineering tool. These equations can be used to embed design intelligence directly into solid models. In this section, the flange created in Section 4.1 will be modified to include parametric equations relating the hole pattern to the rib pattern.

FIGURE 4.52

With the flange model open, select Tools: Equations from the main menu, as shown in Figure 4.52.

The Equations dialog box will appear.

In the Equations dialog box, click the Add button, as shown in Figure 4.53.

The New Equation dialog box will appear, as shown in Figure 4.54. This dialog box allows for equations relating model parameters to one another to be added to the model.

Double-click on the Hole Pattern in the Feature-Manager.

This will display the parameters (6 holes, 360° spacing) in the model window.

FIGURE 4.53

FIGURE 4.54

FIGURE 4.55

Click on the "6" parameter (denoting the number of instances) in the model window, as shown in Figure 4.55.

Note that the symbol denoting this parameter ("D1@Hole Pattern") now appears in the New Equation dialog box. We can now establish a parametric relationship driving this parameter (the number of holes in the pattern) with another parameter in the model. When considering our design intent, the number of holes in the pattern is not necessarily an independent design choice; most likely, we simply want to ensure that there is one hole centered between each pair of

FIGURE 4.56

stiffening ribs. Therefore, we will establish a parametric equation tying the number of holes in the pattern directly to the number of ribs in the Rib Pattern.

Click on the equal sign in the equation box, as shown in Figure 4.56.

Since the parameter representing the number of holes in the Hole Pattern is on the left side of this equal sign, the number of holes is a *driven* parameter; whatever equation is entered on the right side of the equal sign will determine its value. Note that since this value is now a driven parameter, it cannot be directly modified anymore.

The right side of the equation will now be added.

FIGURE 4.57

Click on the check at the top of the Dimension PropertyManager, as shown in Figure 4.57, to close it and restore the FeatureManager. Now double-click on the Rib Pattern to display the parameters associated with it. Click on the "6," as shown in Figure 4.58, to complete the equation.

The equation is shown in Figure 4.59.

FIGURE 4.58

FIGURE 4.59

FIGURE 4.60

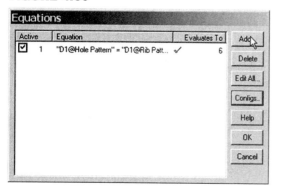

Select OK to add the equation. The Equation dialog box now shows the equation and the value that is calculated for the equation (6), as shown in Figure 4.60. Click OK to close the dialog box.

This equation establishes a parametric relation that ties the number of holes in the hole pattern directly to the number of ribs in the rib pattern. We will test the equation at this point.

With the Rib Pattern parameters still displayed in the model window, double-click on the value "6" (the number of ribs in the pattern) and change it to "3." Click the check mark, and then click the Rebuild Tool to rebuild the model.

FIGURE 4.61

The model will be rebuilt, with the Rib Pattern modified and the driven Hole Pattern modified as well, as shown in Figure 4.61.

While the number of holes is correctly tied to the number of ribs, our design intent may not be satisfied by this model. Note that the angular location of the holes was set at 30 degrees from the center of a rib; this centered the hole between two ribs when there were six ribs in the pattern, but it no longer provides for centering of the holes when the number of ribs is modified. We will create a new parametric equation to establish the relationship required by our design intent; the equations drive the angular dimension of the holes so that they are centered between the ribs.

From the main menu, select Tools: Equations. In the Equation dialog box, click Add. Double-click the Bolt Hole in the FeatureManager to display the dimensions associated with the bolt hole.

The angular spacing between ribs is 360 degrees divided by the number of ribs. The first hole is located at one-half of this value away from the first rib. Therefore, the equation required to set the angular dimension to the desired value will be:

$$\text{Angular Dimension} = \tfrac{1}{2}(360/\text{Number of Ribs}) \text{ or}$$
$$\text{Angular Dimension} = 180/\text{Number of Ribs}$$

With the hole dimensions still displayed, select the angular dimension locating the hole, as shown in Figure 4.62. In the New Equation dialog box, click the "=" sign. Type the number 180, and click the "/" sign.

FIGURE 4.62

In the Dimension PropertyManager, click the check box to return to the FeatureManager. Double-click on the Rib Pattern in the FeatureManager to display the associated parameters. Select the parameter that represents the number of ribs by clicking on the number 3 in the model window, as shown in Figure 4.63.

FIGURE 4.63

FIGURE 4.64

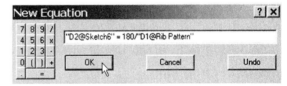

The new equation should be as shown in Figure 4.64.

Click OK to close the New Equation dialog box.

The Equation box confirms that the dimension will equal 60 degrees with the current number of ribs (3), as shown in Figure 4.65.

Click OK to close the Equation box and then rebuild the model.

FIGURE 4.65

FIGURE 4.66

FIGURE 4.67

The result is shown in Figure 4.66.

Change the number of ribs to other values and check to see that the number and locations of the holes change in a consistent manner.

Another example is shown in Figure 4.67.

Note that there is a difference between *driving* and *driven* parameters. In this example, the number of ribs is a free design choice, and is therefore a driving parameter. As such, it appears only on the right side of the equal sign in parametric equations.

Conversely, the number of holes and the angular hole location are *driven* parameters; they are determined by the choice of number of ribs, and appear on the left side of the equal sign in parametric equations. Since their values are set by the values established by the driving parameters, driven parameters cannot be modified directly in the model.

Perform the following demonstration to verify this.

FIGURE 4.68

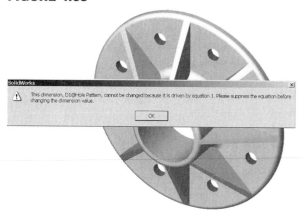

Double-click on the Hole Pattern in the Feature-Manager to display the associated parameters in the model window. Double-click on number of holes in the pattern, to try and change the value.

An error message appears (see Figure 4.68), indicating that the value depends on a driving parameter and cannot be modified. In this way, we have embedded some design intelligence directly into the model.

Click OK to close the message box. Reset the number of ribs and holes to six, and save the part file.

4.3 Modeling Tutorial: Cap Screw with Design Table

Note: This tutorial requires Microsoft Excel to be installed on your computer.

In this section, we will create a family of similar parts. Many parts are defined this way, especially common parts such as fasteners, washers, seal rings, and so on. Rather than creating separate model files and drawings for every different part, a single part with multiple configurations is made. A single drawing can be made to define the parameters of all of the different configurations. The specifications of the dimensions that define each configuration are contained in a spreadsheet called a *design table*.

Sketch

Circle

Smart
Dimension

Open a new part. Select the Right Plane from the FeatureManager, and select the Circle Tool from the Sketch group of the CommandManager. Draw a circle centered at the origin. Select the Smart Dimension Tool, and dimension the diameter of the circle as 0.50 inches, as shown in Figure 4.69.

Select the Extruded Boss/Base Tool from the Features Group of the CommandManager, and extrude the circle 1.50 inches, in the direction shown in Figure 4.70.

Features

Extruded
Boss/Base

FIGURE 4.69

FIGURE 4.70

Direction of Extrusion

FIGURE 4.71

Select the face shown in Figure 4.71.

FIGURE 4.72

Ø.750

Select the Circle Tool from the Sketch group of the CommandManager. Draw a circle centered at the origin. Select the Smart Dimension Tool and dimension the diameter of the circle as 0.75 inches, as shown in Figure 4.72.

Sketch   Circle   Smart Dimension

FIGURE 4.73

Direction of Extrusion

Select the Extruded Boss/Base Tool from the Features Group of the CommandManager, and extrude the circle 0.50 inches to form the head of the screw, as shown in Figure 4.73.

Features   Extruded Boss/Base

Select the top of the screw head, as shown in Figure 4.74.

From the main menu, select Tools: Sketch Entities: Polygon, as shown in Figure 4.75. Drag out a polygon from the origin, as shown in Figure 4.76. By default, the number of sides is six (a hexagon).

Press the esc key to turn off the Polygon Tool, and click on one of the sides of the hexagon to select it. In the PropertyManager, add a Horizontal relation to the side.

FIGURE 4.74

FIGURE 4.75

FIGURE 4.76

FIGURE 4.77

FIGURE 4.78

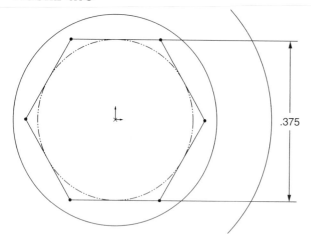

.375

The hexagon will appear as in Figure 4.77.

Select the Smart Dimension Tool from the Sketch group of the CommandManager, and add a 0.375-inch dimension between two opposite sides of the hexagon, as shown in Figure 4.78.

Sketch

Smart Dimension

FIGURE 4.79

The sketch is now fully defined.

Select the Extruded Cut Tool from the Features group of the CommandManager, and extrude a cut 0.245 inches deep to form the hex socket in the head, as shown in Figure 4.79.

Features

Extruded Cut

FIGURE 4.80

Screw threads are rarely modeled as solid features, since they can be cumbersome to create and can significantly slow down performance of your computer system. Also, since they are standardized, the thread profiles are not required in order to specify them on a drawing. Rather, *cosmetic threads*, graphical features used to show the threaded regions, are much more commonly used. Cosmetic threads can be added to any cylindrical feature.

Select the edge where the threads will begin, as shown in Figure 4.80.

Select Insert: Annotation: Cosmetic Thread from the main menu, as shown in Figure 4.81.

FIGURE 4.81

FIGURE 4.82

Set the thread length as 1.00 inch, and the minor diameter as 0.40 inches, as shown in Figure 4.82. Click the check mark to add the thread. The resulting thread display is shown in Figure 4.83.

The minor diameter is the diameter at the "root" of the thread. The actual minor diameter value of a UNF (unified series fine) 1/2-inch thread is 0.408 inches. Since the minor diameter of the cosmetic thread is only for display purposes, using an approximate value is acceptable.

If preferred, the cosmetic thread can be viewed only with dashed lines rather than with the shaded display shown in Figure 4.83. If you want to change the display mode of the cosmetic thread, choose Tools: Options from the main menu, and under the Document Properties tab, click on Annotations Display. Click on the "Shaded cosmetic threads" box, as shown in Figure 4.84. The resulting display, shown in the Front View, is shown in Figure 4.85.

The definition of the thread appears in the FeatureManager, attached to the cylindrical feature that is to be threaded, as shown in Figure 4.86.

FIGURE 4.83

FIGURE 4.84

FIGURE 4.85

FIGURE 4.86

FIGURE 4.87

Save the part file.

We have now defined one configuration of the cap screw. To define more configurations, we are going to create a design table.

We will need to access all of the dimensions used to create the part.

FIGURE 4.88

To show the dimensions, right-click on Annotations in the FeatureManager, and select Show Feature Dimensions, as shown in Figure 4.87.

Move the dimensions around the screen so that they are all easily visible, as shown in Figure 4.88.

From the main menu, select Insert: Design Table, as shown in Figure 4.89

FIGURE 4.89

FIGURE 4.90

In the Property Manager, select Blank as the type. Under Edit Control, select the second option, so that dimensions defined in the table cannot be edited outside of the table. Clear all of the options for automatically adding new rows and columns, as shown in Figure 4.90. Click the check mark to begin creation of the design table.

A window containing a Microsoft Excel spreadsheet appears in the modeling area, as shown in Figure 4.91. (Note: Be careful not to click in the white area of the modeling area while editing the design table. Doing so will close the design table. If you accidentally close the design table before you are finished editing, right-click on the design table in the FeatureManager, and select Edit Table from the menu.)

The first row contains the title of the table. In the second row, beginning with cell B2, the parameters to be specified by the table will be identified.

With cell B2 selected, double-click the 0.50-inch dimension defining the diameter of the shank of the screw, as shown in Figure 4.92.

Note that the SolidWorks name for this dimension, "D1@Sketch1," is shown in cell B2, as shown in Figure 4.93. Also, the value for this dimension is shown in cell B3.

FIGURE 4.91

FIGURE 4.92

FIGURE 4.93

With cell C2 highlighted, double-click the 1.5-inch dimension defining the length of the bolt. Repeat for these other dimensions, as shown in Figure 4.94:

FIGURE 4.94

	A	B	C	D	E	F	G	H
1	Design Table for: CapScrew							
2		D1@Sketch1	D1@Extrude1	D1@Sketch2	D1@Extrude2	D1@Sketch3	D1@Cut-Extrude1	D1@Cosmetic Thread1
3	First Instance	0.5	1.5	0.75	0.5	0.375	0.245	1
4								
5								
6								
7								
8								
9								

D2: Head diameter (0.75 in.)

E2: Head height (0.50 in.)

F2: Hex width (0.375 in.)

G2: Hex depth (0.245 in.)

H2: Thread length (1.00 in.)

FIGURE 4.95

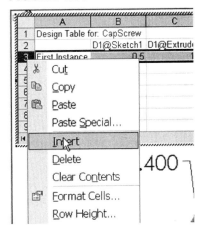

The dimension names assigned by the program are not descriptive of the functions of the dimensions. We can add descriptive names in a new row (these descriptive names will be especially helpful if we include the design table in a drawing, as we will do later in this tutorial).

Right-click on the "3" defining the row number of the spreadsheet, and select Insert from the menu, as shown in Figure 4.95.

In the new row, type the descriptive name in each column, as shown in Figure 4.96. Be sure to leave cell A3 blank.

FIGURE 4.96

	A	B	C	D	E	F	G	H	I
1	Design Table for: CapScrew								
2		D1@Sketch1	D1@Extrude1	D1@Sketch2	D1@Extrude2	D1@Sketch3	D1@Cut-Extrude1	D1@Cosmetic Thread1	
3		Diameter	Length	Head Diamete	Head Height	Hex Width	Hex Depth	Thread Length	
4	First Instance	0.5	1.5	0.75	0.5	0.375	0.245	1	
5									
6									

FIGURE 4.97

	A	B	
1	Design Table for: CapScrew		
2		D1@Sketch1	D1@E
3		Diameter	Lengtl
4	Part 101	0.5	
5			
6	✛		
7			
8			
9			

⊪ ◀ ▶ ⊪ \ Sheet1 /

Below the row containing the SolidWorks dimension names, Column A is reserved for the name of each part configuration.

Click cell A4, and type the name "Part 101," as shown in Figure 4.97. Enter "Part 102" and "Part 103" in cells A5 and A6, respectively. Enter the values shown in Figure 4.98 as the dimensions for the two new configurations.

Click in the white space in the modeling window, outside of the spreadsheet window. This will cause the design table to close, and a message will be displayed that indicates the new configurations have been created, as shown in Figure 4.99. Click OK.

FIGURE 4.98

	A	B	C	D	E	F	G	H	
1	Design Table for: CapScrew								
2		D1@Sketch1	D1@Extrude1	D1@Sketch2	D1@Extrude2	D1@Sketch3	D1@Cut-Extrude1	D1@Cosmetic Thread1	
3		Diameter	Length	Head Diamete	Head Height	Hex Width	Hex Depth	Thread Length	
4	Part 101	0.5	1.5	0.75	0.5	0.375	0.245	1	
5	Part 102	0.75	2	1.125	0.75	0.635	0.37	1.5	
6	Part 103	1	3	1.5	1	0.75	0.495	2	
7									
8									
9									

FIGURE 4.99

Note that the part has not changed on the screen. That is because the default configuration is the one with the dimensions used to model the part originally. To view the new configurations created from the design table, we need to use the ConfigurationManager.

At the top of the FeatureManager, there are tabs corresponding to the FeatureManager, PropertyManager, and ConfigurationManager. (There could be other tabs as well, if certain add-ins are present.) Click on the icon representing the ConfigurationManager, as shown in Figure 4.100.

FIGURE 4.100

In the ConfigurationManager, double-click on Part 103, as shown in Figure 4.101.

The part is rebuilt to the dimensions specified in the design table for Part 103, as shown in Figure 4.102. Dimensions controlled by the design table may appear in a different color on your screen. We will see how to define the color for these dimensions later in this tutorial.

FIGURE 4.101

FIGURE 4.103

FIGURE 4.102

One dimension that was left unchanged is the minor thread diameter of the cosmetic thread. Since this is not a dimension used to define the part, we did not include it in the design table. However, we would like to have the display of the cosmetic thread look reasonable on the screen. We can add an equation to control this dimension.

From the main menu, select Tools: Equations, as shown in Figure 4.103.

Click Add to create a new equation. Click on the dimension representing the minor thread diameter (0.400). In the equation box, type "=0.8*" after the name of the minor diameter dimension ("D2@Cosmetic Thread1") and click on the dimension representing the diameter of the bold shank (ϕ1.00). The equation should appear as shown in Figure 4.104. Click OK to add this equation, and OK again to close the Equation box.

Click the Rebuild Tool.

The updated thread diameter is shown in Figure 4.105.

Click on the FeatureManager tab to return to the FeatureManager. Right-click Annotations in the FeatureManager, and select Show Feature Dimensions, as shown in Figure 4.106, to turn off the display of the dimensions on the screen. Save the part file.

FIGURE 4.104

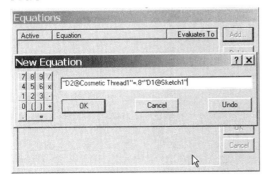

FIGURE 4.105

∅.800

2.000

FIGURE 4.106

4.4 Incorporating a Design Table in a Drawing

We now will make a single drawing that details all three configurations of the cap screw.

Open a new drawing. Choose an A-size landscape sheet size, and uncheck the "Display sheet format" box. If Model View appears in the PropertyManager, uncheck the box labeled "Start command when creating new drawing" and click the × to close the Model View command.

We will use a drag-and-drop technique to import the model into the drawing.

FIGURE 4.107

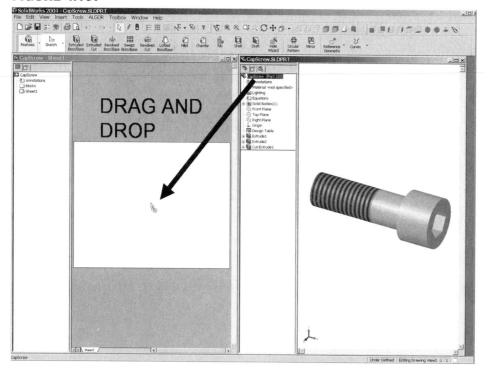

Open the part file, if it is not already open. From the main menu, select Windows:Tile Vertically.

This will allow all open SolidWorks documents to be viewed simultaneously.

In the part window, click on the name at the top of the FeatureManger, and drag the name into the white space of the drawing menu, as shown in Figure 4.107. Three standard views will be created in the drawing window.

The model views that are created reflect the configuration that is current when the drawing is created. The configuration used for any model view can be changed by right-clicking on that view and selecting Properties from the menu that appears. For this tutorial, the configuration used is not important.

FIGURE 4.108

Maximize the drawing window. Select the top view, as shown in Figure 4.108, and press the delete key.

All dimensions defining the part can be shown on the front and right views.

With one of the drawing views selected, select Insert: Tables: Design Table from the main menu, as shown in Figure 4.109.

The design table is inserted into the drawing, as shown in Figure 4.110. The design table can be moved on the drawing sheet by clicking and dragging it, but the format of the

FIGURE 4.109

FIGURE 4.110

Design Table for: CapScrew							
	D1@Sketch1	D1@Extrude1	D1@Sketch2	D1@Extrude2	D1@Sketch3	D1@Cut-Extrude1	D1@Cosmetic Thread1
	Diameter	Length	Head Diameter	Head Height	Hex Width	Hex Depth	Thread Length
Part 101	0.5	1.5	0.75	0.5	0.375	0.245	1
Part 102	0.75	2	1.125	0.75	0.635	0.37	1.5
Part 103	1	3	1.5	1	0.75	0.495	2

table is not what we want. There are blank rows and columns visible, dimensions are shown to only the number of decimal places we input, and so on. We will now edit the table to improve its appearance.

Double-click on the design table, and you will be taken back to the part screen, with the spreadsheet open in a window, as shown in Figure 4.111.

FIGURE 4.111

	A	B	C	D	E	F	G	H	I	J	K
1	Design Table for: CapScrew										
2		D1@Sketch1	D1@Extrude1	D1@Sketch2	D1@Extrude2	D1@Sketch3	D1@Cut-Extrude1	D1@Cosmetic Thread1			
3		Diameter	Length	Head Diamete	Head Height	Hex Width	Hex Depth	Thread Length			
4	Part 101	0.5	1.5	0.75	0.5	0.375	0.245		1		
5	Part 102	0.75	2	1.125	0.75	0.635	0.37		1.5		
6	Part 103	1	3	1.5	1	0.75	0.495		2		
7											
8											
9											

Sheet1

FIGURE 4.112

Remember that the design table is a Microsoft Excel spreadsheet, and we will be editing it the same way we would edit any other Excel spreadsheet.

To begin, the first two rows do not need to be displayed on the drawing. We cannot delete them, since they contain information necessary to the design table, but we can hide them.

Click and hold the mouse key on the number 1 on the left side of the spreadsheet, and drag the cursor down onto the 2 and release. Rows 1 and 2 will be highlighted. Right-click, and select Hide from the menu, as shown in Figure 4.112.

Select Columns B–H. Click the Center icon on the Excel menu, as shown in Figure 4.113.

FIGURE 4.113

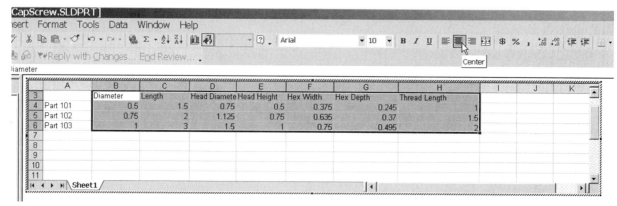

Click on the blank cell in the upper-left corner, which selects the entire spreadsheet. Increase the font size to 14 pt from the pull-down menu, as shown in Figure 4.114.

Adjust the column width of each column by clicking and dragging on the line between two column numbers, as shown in Figure 4.115.

FIGURE 4.114

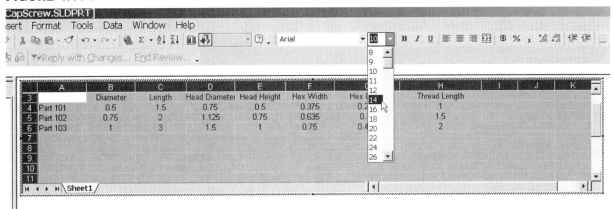

FIGURE 4.115

C	D	E	
ength	ad Diame	ead Heig	He
1.5	0.75	0.5	(
2	1.125	0.75	(
3	1.5	1	

Select the cells con-
taining numbers by
clicking and dragging
across the range of
cells. Right-click and
choose Format Cells
from the menu. Under
the Number tab, set
the number of decimal
places to 3, as shown in Figure 4.116.

FIGURE 4.116

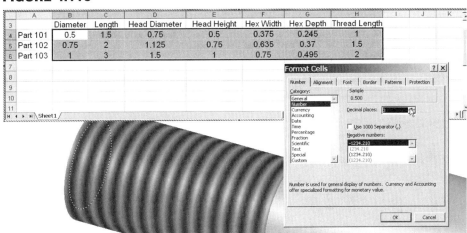

Finally, move the handles on the bottom and right of the spreadsheet so that only cells
with values are displayed, as shown in Figure 4.117.

FIGURE 4.117

	A	B	C	D	E	F	G	H
3		Diameter	Length	Head Diameter	Head Height	Hex Width	Hex Depth	Thread Length
4	Part 101	0.500	1.500	0.750	0.500	0.375	0.245	1.000
5	Part 102	0.750	2.000	1.125	0.750	0.635	0.370	1.500
6	Part 103	1.000	3.000	1.500	1.000	0.750	0.495	2.000

Click outside of the spreadsheet to close it. From the main menu, select Windows and
click on the drawing name to return to the drawing.

Click on the Rebuild Tool to update the design table in the drawing, which now
appears as shown in Figure 4.118.

We will now insert the dimensions from the part into the drawing.

FIGURE 4.118

	Diameter	Length	Head Diameter	Head Height	Hex Width	Hex Depth	Thread Length
Part 101	0.500	1.500	0.750	0.500	0.375	0.245	1.000
Part 102	0.750	2.000	1.125	0.750	0.635	0.370	1.500
Part 103	1.000	3.000	1.500	1.000	0.750	0.495	2.000

FIGURE 4.119

Select the front view, and choose Insert: Model Items from the main menu. In the dialog box, check the box labeled "Include items from hidden features," as shown in Figure 4.119. Click OK.

The depth of the hex cavity could be shown in a section view, but this not necessary for this simple part. Rather, we will show and dimension the hex cavity in the front view.

Select the Hidden Lines Visible Tool from the View toolbar.

FIGURE 4.120

The lines defining the hex cavity are now displayed in the front view, as shown in Figure 4.120. The dimensions defined by the design table may be shown in a different color than black. (The default color is magenta.) While this may be desirable in the part file, for the drawing we would prefer that all dimensions be displayed in black.

Select Tools: Options from the main menu. Under the System Options tab, select colors and scroll down the list of entities to find "Dimensions, Controlled by Design Table," as shown in Figure 4.121. Select Edit, and set the color to black.

FIGURE 4.121

FIGURE 4.122

FIGURE 4.123

D1@Extrude1 of CapScrew

FIGURE 4.124

Move the dimensions and/or views on the screen so that they are all visible. Move the head diameter and hex width dimensions to the right view, as shown in Figure 4.122.

(Recall that to move a dimension from one view to another, click and drag it while holding down the shift key.)

The dimensions displayed are those for one configuration. To make this drawing one that defines the dimensions for all configurations, we need to display the dimension names in the drawing views.

Click on the dimension defining the length, as shown in Figure 4.123. In the PropertyManager, replace the "<DIM>" in the Dimension Text box and type in "Length," as shown in Figure 4.124.

Repeat for the other dimensions. Add notes, if desired, to complete the drawing, as shown in Figure 4.125.

FIGURE 4.125

	Diameter	Length	Head Diameter	Head Height	Hex Width	Hex Depth	Thread Length
Part 101	0.500	1.500	0.750	0.500	0.375	0.245	1.000
Part 102	0.750	2.000	1.125	0.750	0.635	0.370	1.500
Part 103	1.000	3.000	1.500	1.000	0.750	0.495	2.000

Socket Head Cap Screw

Problems

P4.1   Consider the flange model developed in Chapter 1. Create a "blank" for this part
(without the holes) as shown in Figure P4.1a, using the following procedure:

    a.   Sketch a 5.5 inch diameter circle on the Top Plane, and extrude it upward
2.25 inches to create a solid cylinder (Figure P4.1b).

    b.   Using the Convert Entities Tool and appropriate dimensions, sketch a
"cutting tool" to create the flange "blank" from the solid (Figure P4.1c).

    c.   Use the Cut: Revolve command to create the blank.

FIGURE P4.1a

FIGURE P4.1b

FIGURE P4.1c

P4.2   Using equations, create a solid box where the width w is the driving dimension,
the depth $d = 2\,w$, and the height $h = 3w$. Show the box for a few different
values of the driving dimension.

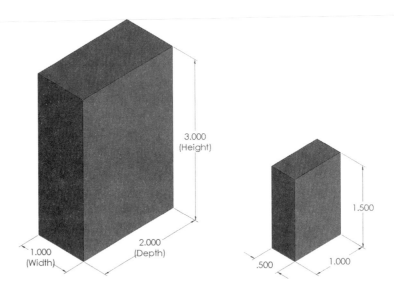

P4.3   Model the bracket shown in Figure P4.3. Use only the dimensions shown here;
use relations and symmetry as required so that these dimensions completely
define the part. (The fillet radius, 0.125 inches, is the same for the three fillets.)

Add equations to the drawing so that these relationships exist between the
dimensions:

FIGURE P4.3

1. Height = 0.40 × Width
2. Leg Width = 1/2 Height
3. Hole Spacing = Width minus 2 inches
4. Hole Location = 1/2 (Height minus 0.25 inches)
5. Slot Width = 1/2 Hole Spacing
6. Slot Location = 1/2 Leg Width

Check to see that the equations work for Width values from 3 to 8 inches.

P4.4 In this exercise, you will model a part in which an integer design parameter (number of holes) is controlled by an equation.

FIGURE P4.4a

FIGURE P4.4b

FIGURE P4.4c

FIGURE P4.4d

FIGURE P4.4e

FIGURE P4.4f

a. Model the part shown in Figure P4.4a. Use a linear pattern to place the second hole. Show all of the dimensions by right-clicking on Annotations in the FeatureManager and clicking on "Show Feature Dimensions."

b. Add an equation so that the number of holes is equal to the length of the part (4 inches in the current configuration) divided by two. Therefore, there will be one hole for every 2 inches of length. Show only the dimensions shown in Figure P4.4b by right-clicking on each of the other dimensions and selecting "Hide."

c. Change the length of the part to 5 inches, as shown in Figure P4.4c. Note that the program has rounded the value of the equation (2.5) to the closest integer value (3).

d. Add a second equation to change the value of the dimension specifying the location of the first hole so that the holes will be centered on the part, as shown in Figure P4.4d.

Experiment with several values of length to show that the equations produce the desired results, as shown in Figures P4.4e and P4.4f.

P4.5   Use a design table to create the W18 series of wide-flange shape beams, according to Figure P4.5 and Table P4.5. The model of the beam should be extruded to 36 inches in all configurations.

FIGURE P4.5

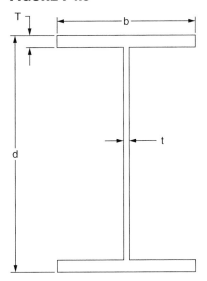

Table P4.5

Designation	Depth *d*	Flange width *b*	Flange thickness *T*	Web thickness *t*
W18 × 106	18.73	11.200	0.940	0.590
W18 × 76	18.21	11.035	0.680	0.425
W18 × 50	17.99	7.495	0.570	0.355
W18 × 35	17.70	6.000	0.425	0.300

Note: All dimensions in inches.

P4.6   Create a multiconfiguration drawing of the model created in P4.5.

P4.7   Make a copy of the flange created in Chapter 1, which is shown in Figure P4.7a with some of the dimensions hidden. Create three different configurations of the flange (the flange as created in Chapter 1 will be the first configuration), using both equations and a design table, as detailed below.
a.   Add two equations:
   (1)   The boss diameter (2.75 in.) is equal to one-half of the flange diameter (5.50 in.).
   (2).   The center hole diameter (1.50 in.) is equal to the boss diameter minus 1.25 inches.
b.   Create a design table to define the dimensions of two additional configurations, as specified in Table P4.7.

Table P4.7

Flange	Diameter	Height	Bolt circle diameter	Number of bolt holes
Part 1	5.5	2.25	4.25	4
Part 2	7.0	3.00	5.25	6
Part 3	8.0	3.50	6.25	8

Note: All dimensions in inches.

FIGURE P7.a

FIGURE P7.b

Part 1   Part 2   Part 3

c.   Make a 2-D drawing showing the three configurations, with only the
      dimensions that change shown, as in Figure P4.7b. This type of drawing is
      used often in product literature and catalogs to illustrate the relative sizes of
      different parts.

      Insert three front views and three top views in the drawing. Right-click
      on each view and select Properties, and select the appropriate named
      configuration for each view. Hide unwanted dimensions by selecting View:

Hide/Show Annotations from the main menu and selecting the dimensions to be hidden. Press the esc key to return to the normal drawing mode.

There are two ways to align the drawing views with each other. To precisely align the views, select a view, right-click, and choose Align: Align Horizontal by Origin, and select another view to align to. Repeat until all views are aligned. If only approximate alignment is needed, turn on the grid display from Tools: Options and use the gridlines to align the views.

Advanced Concepts in Part Modeling

Introduction

The parts that we have made so far have been made primarily with extruded and revolved bases, bosses, and cuts. In this chapter, we will introduce several other tools for creating and modifying parts, including the *Loft, Sweep,* and *Shell* Tools.

In this chapter, you will:

- create a lofted feature from multiple sketches,
- use the Shell Tool to create a thin-wall part,
- learn how to change the order in which features are created and modified,
- create raised letters on a part, and
- create a feature by sweeping a cross-section around a planar or 3-D path.

5.1 A Lofted and Shelled Part

In this exercise, we will construct the business card holder shown in Figure 5.1. Note that the top of the part is rounded, while the bottom of the part is rectangular. These dissimilar shapes will be joined into a solid with the Loft Tool. Also, notice that the part is not solid, but rather is hollow underneath, as the view in Figure 5.2 shows. The Shell Tool allows this type of construction to be easily modeled.

FIGURE 5.1   **FIGURE 5.2**

Open a new part. Select the Top Plane, and then select the Rectangle Tool from the Sketch group of the Command-Manager. Drag out a rectangle.

Sketch   Rectangle

Select the Centerline Tool, and add a centerline from one corner of the rectangle to the opposite corner. Press the esc key to turn off the Centerline Tool.

Centerline

Select the centerline and the origin, and add a Midpoint relation, centering the rectangle about the origin.

Smart Dimension

Select the Smart Dimension Tool, and dimension the rectangle to be 5.50 inches by 2.00 inches, as shown in Figure 5.3.

In previous tutorials, upon completing a sketch we have then used a features tool to convert the sketch into a 3-D object. However, a lofted feature requires at least two sketches. Therefore, we will close this sketch and begin the second sketch.

FIGURE 5.4

FIGURE 5.3

Close the sketch by clicking on the icon indicated in Figure 5.4, in the upper-right corner of the screen.

The second sketch, which will define the top of the part, will be created in a new plane.

FIGURE 5.5

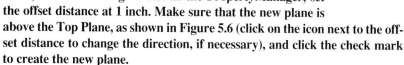

Select the Top Plane from the FeatureManager. Select the Features group of the CommandManager. Click on the Reference Geometry Tool, and choose Plane from the options listed, as shown in Figure 5.5. In the PropertyManager, set the offset distance at 1 inch. Make sure that the new plane is above the Top Plane, as shown in Figure 5.6 (click on the icon next to the offset distance to change the direction, if necessary), and click the check mark to create the new plane.

Features

Select the Line Tool from the Sketch Group of the CommandManager.

Sketch   Line

Draw two horizontal lines, as shown in Figure 5.7, and add an Equal relation to them. Also add a vertical relation between two corresponding endpoints.

FIGURE 5.6

Select the Tangent Arc Tool. Drag an arc between corresponding endpoints of the two lines. Be sure to snap to the endpoint of the line for both arc endpoints, as shown in Figure 5.8.

FIGURE 5.7

Plane1

Repeat for the other ends of the lines. Select the Centerline Tool, and add a diagonal centerline. Add a Midpoint relation between the centerline and the origin. Select the Smart Dimension Tool, and add the dimensions shown in Figure 5.9 to fully define the sketch. Close the sketch.

FIGURE 5.8

A = 180°
R = 0.49

Both of the sketches required for the lofted feature are now in place.

Select the Lofted Boss/Base Tool from the Features group of the CommandManager. Now click on each of the two sketches, near corresponding corners, as shown in Figure 5.10.

The lofted feature will be created based on a "guide curve." The guide curve is created based on the location of the selection of the sketches, as previewed in Figure 5.11. Click the check mark to complete the loft. The resulting part is shown in Figure 5.12.

Features   Lofted Boss/Base

FIGURE 5.9

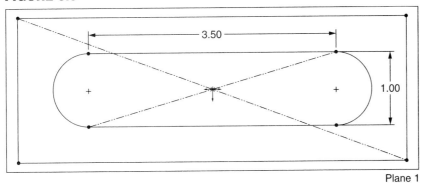

3.50

1.00

Plane 1

FIGURE 5.10

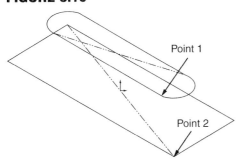

Point 1

Point 2

FIGURE 5.11

FIGURE 5.12

Select the top surface of the part. Select the Sketch Tool from the Sketch group of the CommandManager. Select all of the edges of the top surface (the two lines and two arcs of the sketch, and select the Offset Entities Tool.

A preview of the offset operation will be displayed, as shown in Figure 5.13.

FIGURE 5.13

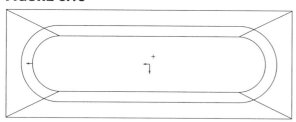

In the PropertyManager, set the offset distance as 0.20 inches in the dialog box. Check the Reverse box, as shown in Figure 5.14, so that the offset entities are inside of the edges of the face, as shown in Figure 5.15. Click on the check mark to finish.

The finished sketch is shown in Figure 5.16. Note that the 0.20-inch offset distance is shown as a dimension that can be edited by double-clicking it.

FIGURE 5.14   **FIGURE 5.15**

FIGURE 5.16

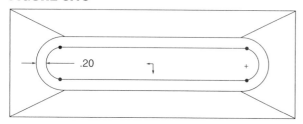

Before cutting the shape of the sketch into the part, we need to consider the design intent. We want this cut to be blind, so that the cards sit on the bottom of the hole, but we want the cut depth to vary with the overall height of the card holder. Therefore, neither a Blind nor Through All type of cut will work. Rather, we will specify the depth of cut so that the bottom of the cut is a fixed distance above the bottom of the card holder.

FIGURE 5.17

Select the Extruded Cut Tool from the Features group of the CommandManager. Switch to the Trimetric View. In the PropertyManager, set the type of extrusion to Offset from Surface, as shown in Figure 5.17. As the surface, we want to select the bottom face of the part.

To select the bottom surface, we could switch to Bottom view or rotate the model until the bottom view is visible, but in the steps that follow, we will learn a handy technique for selecting a nonvisible surface *without* rotating the model from the trimetric view.

Move the mouse above the bottom face. Do not click the left button; doing so would select the visible outer surface. Right-click, and choose Select Other from the menu, as shown in Figure 5.18.

With the bottom face highlighted, as shown in Figure 5.19, click the left mouse button (Y).

FIGURE 5.18

FIGURE 5.19

FIGURE 5.20

If there were other possible selections available, clicking the right mouse button (N) allows you to scroll between the possible selections.

In the PropertyManager, set the offset distance as 0.125 inches, as shown in Figure 5.20, and click the check mark to complete the cut.

The resulting geometry is shown in Figure 5.21.

We will now introduce the Shell command. This command is often used with molded plastic parts to create thin-wall geometries. Since the part we are modeling does not need to have any significant strength, making the part solid would be a waste of material. We will make the wall thickness of the part a constant 0.060 inches.

FIGURE 5.21

FIGURE 5.23

Select the bottom face of the part as the face to be removed, as shown in Figure 5.22. Select the Shell Tool from the Features group of the CommandManager. In the PropertyManager, set the wall thickness to 0.060 inches, as shown in Figure 5.23. Click the check mark to complete the shell operation, the results of which are shown in Figure 5.24.

FIGURE 5.22

Select the Fillet Tool from the Features group of the CommandManager. In the PropertyManager, set the fillet radius of

FIGURE 5.24

FIGURE 5.25

FIGURE 5.26

FIGURE 5.27

0.10 inches. Select the top face, as shown in Figure 5.25, and click the check mark to apply the fillets.

Note that fillets are usually applied to edges, not faces. Selecting a face causes all of the edges of that face to filleted, as shown in Figure 5.26.

Select the Fillet Tool again. Select one of the edges at the bottom of the cavity (as long as the Tangent propagation box is checked, then the fillet will be extended completely around the bottom edge). Set the radius as 0.050 inches, and click the check mark to apply the fillet, which is shown in Figure 5.27.

Fillet

Although our part appears to be finished, there is a problem that may not be evident from examining the part from standard views. We will use a section view to get a better look at the problem areas.

Select the Front Plane from the FeatureManager. Select the Section View Tool. Click the check box, and the cross-section of the part is displayed, as shown in Figure 5.28.

FIGURE 5.28

Zooming in on the filleted edges shows the problem. Because the filleting was performed after the shell operation, the wall is thinner than desired at the upper corners, as shown in Figure 5.29. (Similarly, the wall is thicker than desired in the lower corners.)

To correct this problem, we could fillet the sharp corners to maintain a constant wall thickness. We could also shell the part *after* creating the fillets. This is the easier way to produce the constant wall thickness. It is not necessary to delete the shell and fillet operations and redo them in the proper order; we can simply reorder them in the FeatureManager.

FIGURE 5.29

FIGURE 5.30

Click on the Shell in the FeatureManager, as shown in Figure 5.30, and hold down the left mouse button.

Drag the cursor until the arrow points below the two fillets, as shown in Figure 5.31.

Release the mouse button, and the features are reordered, as shown in Figure 5.32.

FIGURE 5.31

FIGURE 5.32

FIGURE 5.33

The wall thickness is now constant, as shown in Figure 5.33. Note that not all features can be reordered, as some operations will be based on geometries created by prior operations.

Click on the Section View Tool.

This will toggle off the display of the section view.

Text can be added to a part as a sketch entity and then extruded into raised or embossed letters. We will add a part number in raised letters to the bottom of the part.

 Switch to Bottom View and open a sketch on the flat surface shown in Figure 5.34 by selecting the Sketch Tool from the Sketch group of the CommandManager.

Sketch   Sketch

FIGURE 5.35

FIGURE 5.34

From the main menu, select Tools: Sketch Entities: Text, as shown in Figure 5.35. In the PropertyManager, type in the text, as shown in Figure 5.36 (it is not necessary to select a curve to align the text along).

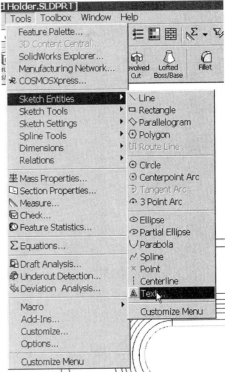

By default, the font specified in the Options will be used. To use another font, click on the "Use document's font" box to uncheck it, which will allow you to select the Font button and edit the type and size of the font.

Click in the sketch at the approximate location of the text, and then click the check mark in the PropertyManager to close the text box.

The text can now be moved by clicking and dragging it around the sketch. Center the text on the face, as shown in Figure 5.37.

If necessary, dimensions can be added to the marker at the lower left of the text to positively locate the text.

FIGURE 5.36

Text here

FIGURE 5.37

FIGURE 5.38

Features

Extruded
Boss/Base

Select the Extruded Boss/Base Tool from the Features group of the Command-Manager.

Extrude the sketch 0.020 inches out from the part, as shown in Figure 5.38.

This card holder may be made from a translucent plastic. The SolidWorks program allows parts to be displayed in the color desired, and also for optical properties such as transparency and shininess to be set.

Select the part name from the FeatureManager (so that the entire part is selected), and select the Edit Color Tool.

In the PropertyManager, select a gray color, and move the transparency slider bar toward the right, as shown in Figure 5.39. Click the check mark to apply the desired color and properties.

FIGURE 5.39

FIGURE 5.40

FIGURE 5.41

FIGURE 5.42

FIGURE 5.43

FIGURE 5.44

The translucent part is shown in Figure 5.40.

Save this part for use in future exercises.

Note that more than two sketches can be used to create a lofted feature. Consider the three sketches shown in Figure 5.41.

The lofted solid created from these sketches is shown in Figure 5.42.

Also, an additional sketch defining a *guide curve* can be introduced, allowing more control over the loft. Figure 5.43 shows a guide curve added to the three previous sketches.

The resulting solid is shown in Figure 5.44.

FUTURE STUDY

Industrial Design

In this chapter, we created a part by using the Loft command to smoothly blend two shapes. Engineers in many industries are sometimes reluctant to use this type of construction, since the resulting surfaces are difficult to define mathematically. This makes the geometry difficult to specify on engineering drawings and often impossible to make with traditional manufacturing methods.

Complex geometries have always been an important tool for *industrial designers*. Industrial designers are important members of product development teams, but perform a different task than do design engineers. The Industrial Designers Society of America defines the role of the industrial designer as:

> The industrial designer's unique contribution places emphasis on those aspects of the product or system that relate most directly to human characteristics, needs and interests. This contribution requires specialized under-

standing of visual, tactile, safety and convenience criteria, with concern for the user. (www.idsa.org)

The image that many associate with industrial designers is that of an artist working on a clay model of an automobile, creating the shapes that would eventually be seen on the showroom floor. The clay model would eventually be digitized by measuring thousands of points on the surface in order to define the shape for the tooling used to stamp the sheet metal body parts. Industrial designers now create many of their models with the "virtual clay" of solid modeling software, computer-controlled machining centers, and rapid prototyping machines.

The roles of design engineer and industrial designer have begun to overlap as they have access to a similar set of product development tools. However, engineers should recognize and utilize the unique capabilities of industrial designers within the product design process.

5.2 A Part Created with a Swept Geometry

A helical spring is a good example of a part created by "sweeping" a cross-section geometry along a path. In this case, a circular cross-section will be swept around a helical path.

FIGURE 5.45

Sketch   Circle   Smart Dimension

Open a new part. Select the Front Plane from the FeatureManager. Select the Circle Tool from the Sketch group of the CommandManager. Draw a circle centered at the origin. Select the Smart Dimension Tool, and dimension the circle's diameter as 2 inches, as shown in Figure 5.45.

From the main menu, select Insert: Curve: Helix/Spiral, as shown in Figure 5.46.

A helix can be defined by specifying any two of the following three quantities:

1. The height, or overall length of the helix,
2. The pitch, the distance between similar points on successive turns of the helix, and
3. Revolutions, the total number of complete turns of the helix.

We will define the height and the number of revolutions and allow the pitch to be calculated.

FIGURE 5.46

FIGURE 5.47

Select Height and Revolution in the "Defined by" box. Set the height to 6 inches, the number of revolutions to 8, and the starting angle to 180 degrees, as shown in Figure 5.47.

Note that the pitch is calculated as 0.75 inches (6 inches divided by 8). The starting angle is not critical, but setting it at a multiple of 90 degrees will result in a cross-section that is oriented with one of the standard views. The helix can be specified as clockwise or counterclockwise. A preview of the helix geometry is shown in Figure 5.48.

FIGURE 5.48

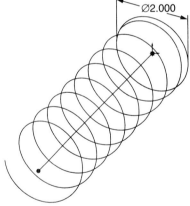

Click OK to accept the definition and close the sketch.

The second sketch necessary to create the part must be created in a plane that is perpendicular to the helix at an endpoint of the helix. By setting the starting angular location of the helix at 180 degrees, the Top Plane is the correct location for the second sketch.

Select the Top Plane from the FeatureManager. Select the Circle Tool. Move the cursor to the endpoint of the helix. Make sure that the box, indicating a snap to a point, is visible next to the cursor before dragging out the circle, as shown in Figure 5.49.

FIGURE 5.49

Drag out a circle. Select the Smart Dimension Tool, and add a 0.25-inch-diameter dimension to the circle, as shown in Figure 5.50. Close the sketch.

Select the Swept Boss/Base Tool from the Features group of the

CommandManager. Select the sketch defining the cross-section (the 0.25-inch-diameter circle) as the Profile and the helix as the Path, as shown in Figure 5.51.

Click the check mark to complete the sweep operation.

The finished spring appears in Figure 5.52.

FIGURE 5.50

Ø.250

FIGURE 5.51

FIGURE 5.52

5.3 A Part Created with a 3-D Sketch as the Sweep Path

The use of a 3-D sketch as a sweep path allows more complex parts to be created. As the name implies, a 3-D sketch contains entities in 3-D space, whereas typical sketches contain entities that exist in a plane. Not all sketch entities are available in a 3-D sketch. Only lines, centerlines, splines, points, and sketch fillets can be used in 3-D sketches.

In this exercise, we will model the handlebars shown in Figure 5.53. Since the handlebars are defined in metric units, we will need to set the units accordingly.

Open a new part. Select Tool: Options from the main menu. Under the Document Properties tab, select units, and choose MMGS (millimeters, grams, seconds).

From the main menu, select Insert: 3-D Sketch, as shown in Figure 5.54.

When working with 3-D sketches, the axes displayed in the corner of the screen, shown in Figure 5.55, are especially important. When we draw lines, we will do so in one of the primary planes—XY, YZ, or ZX.

Select the Line Tool from the Sketch group of the CommandManager.

Sketch   Line

Note that beside the cursor, the plane that the line will be drawn in is displayed, as shown in

FIGURE 5.53

FIGURE 5.54

FIGURE 5.55

FIGURE 5.56   **FIGURE 5.57**   **FIGURE 5.58**

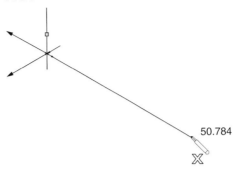

Figure 5.56. We want our first two lines to be sketched in the XZ plane, so we will change this orientation before proceeding.

Press the tab key, which causes the sketch plane to cycle between the three principal planes. Stop when the plane selected is ZX, as shown in Figure 5.57.

Drag out a line from the origin along the X axis, as shown in Figure 5.58. Make sure that the X is displayed before releasing the mouse button.

FIGURE 5.59

FIGURE 5.60

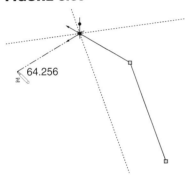

From the endpoint of the first line, drag out a diagonal line, as shown in Figure 5.59.

When dimensioning a 3-D sketch, not all of the options available with 2-D sketches can be used. For example, if we try to add a dimension from the origin to the endpoint of the second line, only the straight-line distance between the two points can be displayed. If we want to dimension the x and y distances from the origin to the point, then the use of a centerline is necessary.

Select the Centerline Tool. Make sure that the ZX designation still shows beside the cursor. Drag out a centerline from the origin along the Z axis, as shown in Figure 5.60.

Select the Smart Dimension Tool. Click once on the first solid line drawn, and place a 50 mm dimension, as shown in Figure 5.61.

Add other dimensions between the last endpoint of the second solid line and the centerline, and between the last endpoint of the second solid line and the first solid line, as shown in Figure 5.62.

The other lines of the sketch will be drawn in the YZ plane.

Select the Line Tool. Press the tab key until YZ is shown as the drawing plane, as shown in Figure 5.63. (If the plane does not change when you press the tab key, then click once on the axis display in the corner to "reset" this feature.)

Drag out a line from the last endpoint of the second solid line in the Z-direction, as shown in Figure 5.64.

Centerline

Smart
Dimension

FIGURE 5.61

FIGURE 5.62

FIGURE 5.63   **FIGURE 5.64**

Drag a line downward (in the –Y direction), as shown in Figure 5.65.

Drag the last line in the –Z direction, as shown in Figure 5.66.

FIGURE 5.65   **FIGURE 5.66**

FIGURE 5.67

Select the Smart Dimension Tool. Add the dimensions shown in Figure 5.67.

Note that the sketch is not fully defined. 3-D sketches are more difficult to fully define than are 2-D sketches. Although a line may be drawn in a specific plane, it is not *locked* into that plane unless it is aligned with one of the principal axes. We could fix some of the endpoints to fully define the sketch, but that is not necessary.

We will now add fillets to the sharp corners of our sketch. The fillet radii to be added are shown in Figure 5.68.

Select the Sketch Fillet Tool. Add the fillets one at a time, as shown in Figure 5.68. The finished sketch is shown in Figure 5.69.

Close the sketch.

FIGURE 5.68

Fillet radius =
100 mm

50

Fillet radius =
30 mm

210

20

70

Fillet radius =
40 mm

140

Fillet radius =
70 mm

170

FIGURE 5.69

50

R100

210

20

R40

R30

70

140

R70

170

Select the **Right Plane** from the **FeatureManager**. Select the **Circle Tool**. Drag out two circles from the origin, as shown in Figure 5.70.

Add diameter dimensions of 25.4 mm and 23.5 mm, as shown in Figure 5.71.

Circle

FIGURE 5.70

FIGURE 5.71

Ø25.400

Ø23.400

(Although most bicycle components are specified in metric units, a 1-inch outer handlebar diameter, 25.4 mm, is a standard size.)

Close the sketch.

Select the **Swept Boss/Base Tool** from the Features group of the CommandManager. In the PropertyManager, select the 2-D sketch just completed as the profile and the 3-D sketch as the path, as shown in Figure 5.72.

Click the check mark to complete the sweep, as shown in Figure 5.73.

FIGURE 5.72

Sweep

Profile and Path

Sketch1

3DSketch1

Features   Swept Boss/Base

FIGURE 5.73

FIGURE 5.74

Select the Mirror Tool. In the PropertyManager, select the Right Plane as the mirror plane, as shown in Figure 5.74, and click the check mark.

The completed handlebars are shown in Figure 5.75.

We will now determine the mass of the handlebars, which will be made from 6061 aluminum.

Right-click on Material in the FeatureManager. Select Edit Material. Select 6061 Alloy from the list of aluminum alloys, as shown in Figure 5.76.

FIGURE 5.75

FIGURE 5.76

FIGURE 5.77

FIGURE 5.78

Select Tools: Mass Properties from the Main Menu, as shown in Figure 5.77.

The mass of the handlebars is calculated at about 221 grams, as shown in Figure 5.78.

Problems

P5.1 Create the solid object shown with a loft between a square base and a circular top (dimensions are in inches).

FIGURE P5.2a

P5.2 Create the part shown in Figure P5.2a. Create the sketch shown in Figure P5.2b in the Top Plane. Create a new plane 4 inches above the Top Plane, and create the second sketch consisting of the two arcs and two lines indicated in Figure P5.2c, snapping to corresponding points of the first sketch. Create a loft between the two sketches to finish the part. All dimensions are in inches.

FIGURE P5.2c

FIGURE P5.2b

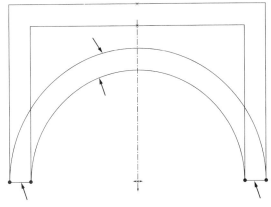

P5.3   Create the part shown here, with a circular cross-section of 1 inch diameter. All dimensions are in inches.

P 5.4   Create the shape shown in Figure P5.4a as a loft defined by the two ellipses and one circle shown in Figure P5.4b. All dimensions are in inches.

FIGURE P5.4a   **FIGURE P5.4b**

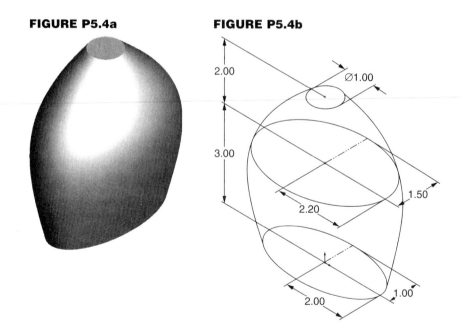

To create an ellipse, choose Tools:Sketch Entity:Ellipse from the main menu. Drag out a circle from the origin, as shown in Figure P5.4c, and then click and drag a point on the edge of the circle to "flatten" it into an ellipse, as shown in Figure P5.4d. Snap points are created at four locations on the ellipse; use these points to define the semimajor and semiminor axes of the ellipse. Add a centerline from the origin to one of the snap points, as shown in Figure P5.4e, and set it to horizontal or vertical to correctly align the ellipse and fully define the sketch.

FIGURE P5.4c

FIGURE P5.4d

R = 2.064, r = 1.121

R = 2.064,
r = 2.064

FIGURE P5.4e

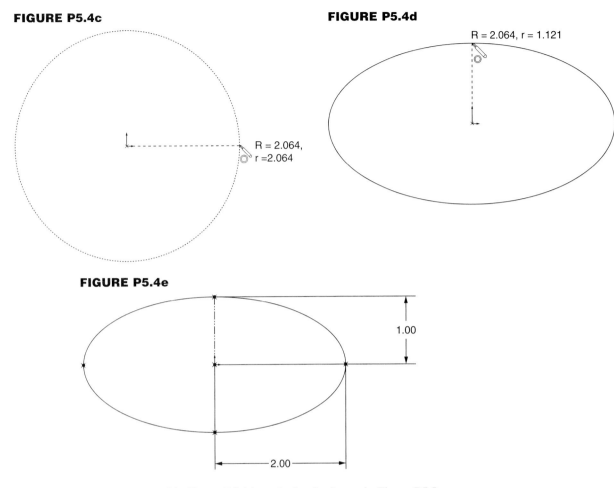

1.00

2.00

P5.5 Turn the shape created in Figure P5.4 into the bottle shown in Figure P5.5a.
Extrude a circular neck, as shown in Figure P5.5b, add a 0.25-inch fillet to the
neck-to-body junction, and use the Shell Tool to hollow the bottle, leaving a
0.020 inch wall thickness. What is the volume of the space within the bottle? Hint:
To find the volume, move the rollback bar to just before
the shell command, as shown in Figure P5.5c and find

FIGURE P5.5a

FIGURE P5.5b

.50

FIGURE P5.5c

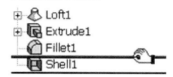

the volume of the solid before shelling. Then move the rollback bar past the shell, and find the volume of the bottle itself. The difference in the two volumes is the volume contained within the bottle.

(Answer: 39.1 cubic inches)

P5.6 Model a bent tube with 0.5-inch outer diameter and 0.375-inch inner diameter to join the ends of the two tubes shown in Figure P5.6a. Follow the route shown by the centerlines in Figure P5.6a, and add 1-inch radius fillets to the corners.

The completed model is shown in Figure P5.6b.

FIGURE P5.6a

FIGURE P5.6b

P5.7 Repeat Problem P5.6, using a different path. Use a tube geometry that minimizes the tube length while maintaining a straight section at each end, as shown in Figure P5.7

FIGURE P5.7

Building Assembly Models from Part Models

Introduction

In the preceding chapters, the development of solid models of parts was covered in detail. In this chapter, methods for combining such part models into complex, interconnected solid models will be described. These types of models, composed of interconnected part models, are called *assembly models*.

The assembly that will be constructed in this chapter is a model of a hinged hatch, as shown in Figure 6.1.

The chapter will begin with a tutorial describing the construction of the part models of four components used in this assembly. After the models are constructed, they will be interconnected first into a simple subassembly, and then into a more complex assembly model. The subassembly and assembly models will be used in subsequent chapters to demonstrate some advanced assembly modeling features.

In this chapter you will:

- create holes using the Hole Wizard,
- learn to import part models into an assembly,
- define assembly mates between parts, and
- create subassemblies, and use them within larger assemblies.

FIGURE 6.1

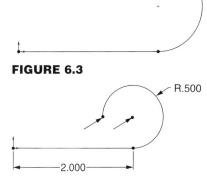

6.1 Creating the Part Models

Before an assembly can be created, the parts to be assembled must be modeled. In this first step, solid models of the four components will be created.

The first model that we will create is the hinge component.

Start a new part model, and sketch a horizontal line from the origin and a tangent arc in the Front Plane, as shown in Figure 6.2.

Add the dimensions shown in Figure 6.3. Add a Horizontal relation between the endpoint of the arc and the center of the arc, as indicated in Figure 6.3.

The sketch should now be fully defined. While in previous exercises we have used sketches with closed contours to make extrusions, in this case we will use the open-contour sketch to create a *thin-feature* extrusion.

FIGURE 6.2

FIGURE 6.3

R.500

2.000

FIGURE 6.4

Create an Extruded Base. Note that the Thin Feature box is checked, since the sketch contour is open. Set the extrusion depth at 4 inches and the thickness at 0.25 inches, as shown in Figure 6.4. Extrude the cross-section 4 inches, resulting in the basic hinge shape. Change the directions as necessary so that the extrusion and thickness directions are as shown in the preview in Figure 6.5.

FIGURE 6.5

The completed extrusion is shown in Figure 6.6.

Select the top horizontal surface of the hinge, as shown in Figure 6.7, for the next sketch.

Go to a top view. Sketch the 1 inch square area to be cut away from the basic hinge shape, as shown in Figure 6.8. Use snaps and/or relations to align the edges of the square with the edges of the hinge, so that the sketch is fully defined with only the single dimension shown.

Use the Extruded Cut Tool to cut Through All in both Direction 1 and Direction 2, as shown in Figure 6.9.

The result of the cut is shown in Figure 6.10.

Select the extruded cut from the FeatureManager, and define a linear pattern to create another instance of this feature 2 inches along the horizontal edge of the hinge, as shown in Figure 6.11.

A pattern of countersunk screw holes will now be added to the hinge. Since fastener holes are generally of standard dimensions, an intelligent design tool known as the Hole Wizard will be used to create the holes.

From the main menu, select Insert: Features: Hole: Wizard, or click on the Hole Wizard Tool from the Features group of the CommandManager.

The Hole Wizard dialog box appears. The Hole Wizard can be used to create holes to accommodate most standard fastener types. We will create countersunk holes for #10 flat head machine screws.

FIGURE 6.6

FIGURE 6.7

FIGURE 6.8

FIGURE 6.9

FIGURE 6.10

FIGURE 6.11

In the Hole Wizard dialog box, select the Countersink tab. Using the pull-down menus, set the parameters for ANSI Inch, Flat Head Screw (100), #10, and Through All, as shown in Figure 6.12. Other parameters can be left at their default values. Click the Next button.

We are now prompted to enter the hole location, as shown in Figure 6.13. We will create a single hole, and replicate it using a linear pattern.

Change to a top view. Place the center of the hole by clicking in the approximate location shown in Figure 6.14.

FIGURE 6.12

FIGURE 6.13

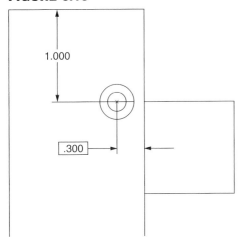

Select the Smart Dimension Tool, and add dimensions to the hole location as shown in Figure 6.15.

Click Finish to create the hole, which is shown in Figure 6.16.

Select the new hole from the FeatureManager, and create a linear pattern creating four total instances of the hole, spaced 2 inches along the long side of the hinge and 0.9 inches along the short side.

The final model of the hinge is shown in Figure 6.17. Since it will be used in a later assembly it must be saved.

Save this using the file name "hinge," and close the file.

FIGURE 6.14

FIGURE 6.15

FIGURE 6.16

Now, the second major component in the first subassembly will be created.

Open a new part, and sketch a 16-inch by 16-inch square in the Front Plane, centered about the origin. Extrude it 2.5 inches.

FIGURE 6.17

FIGURE 6.18

This base feature is shown in **FIGURE 6.19** Figure 6.18.

On the front face, sketch a 10-inch square centered about the origin. Extrude it 1 inch from the face.

This completes the second component, as shown in Figure 6.19. When we assemble the parts later, it will be helpful if they are different colors.

If desired, select the part name from the FeatureManager, and select the Edit Color Tool. Pick the new color for the hatch, and click the check mark to close the Color PropertyManager.

Save it using the file name "hatch." Close the file.

A model of the hinge pin will now be created.

Open a new part, and sketch a 0.5-inch diameter circle centered at the origin. Extrude it 4 inches. Add a 1-inch diameter cap on the pin, extruding it 0.25 inches.

The pin is shown in Figure 6.20. Again, using a new color for this component will aid in visualization of the final assembly.

FIGURE 6.20

Change the color of the part, if desired, and save the part in a file named "pin." Close the file.

The final component needed for the assembly is a frame.

Open a new part, and sketch a 25-inch square centered at the origin. Extrude it to a depth of 4 inches. Extrude a 14-inch square cut (centered on the front face of the component), yielding the part shown in Figure 6.21.

FIGURE 6.21

Change the color, if desired. Save this part in a file named "frame," and close it.

These four parts will now be used to create a model of the hatch assembly.

6.2 Creating a Simple Assembly of Parts

The features of the software that we will employ are the *assembly* capabilities. *Assemblies* are complex solid models that are made up of simpler part models, with specifically defined geometric relationships between the parts. The SolidWorks program provides us with the ability to relate surfaces and other geometric features of one part to those of another part as follows:

- *Define two flat surfaces* as **coincident:** This places the two flat surfaces in the same plane.

- *Define two flat surfaces* as **parallel.**

- *Define two flat surfaces a preset* **distance** *apart:* This makes the two surfaces parallel, with a specified distance between them.

- *Define two lines or planes as* **perpendicular** *to one another.*

- *Define two lines or planes at a preset* **angle** *to one another.*

- *Define a cylindrical feature as* **concentric** *with another cylindrical feature:* This aligns the centerlines of two cylindrical features.

- *Define a cylindrical feature as* **tangent** *to a line or plane.*

FIGURE 6.22

These geometric relationships are called *mates.* There are many other geometric relationships that can be accommodated as well.

In this section, a tutorial will be presented in which we will create a simple assembly by attaching a set of hinges to the hatch component, as shown in Figure 6.22. This assembly will be used in the following section as a small assembly (subassembly) within a larger assembly.

FIGURE 6.23

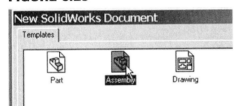

To begin creating an assembly model, select File: New from the main menu. Click on the Assembly icon (Figure 6.23), and click OK.

FIGURE 6.24

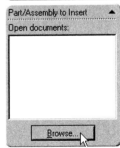

A new assembly window will be created. For the purposes of this assembly, the main "base" part will be the hatch. We will begin by importing this component into the assembly.

In the PropertyManager, any open parts will be displayed. Since we have closed our part files, click on Browse to find the hatch file, as shown in Figure 6.24. (Note: If the box shown does not appear when you start a new assembly, select Insert: Component: Existing Part/Assembly from the main menu.)

Browse to the location where the hatch file was saved, as shown in Figure 6.25, and select it.

FIGURE 6.25

FIGURE 6.26

FIGURE 6.27

The assembly window will return, with a part icon located next to the cursor.

Move the cursor to the origin, and click (Figure 6.26).

This will import the hatch model into the assembly, with the origin of the hatch coincident with the origin of the assembly window.

Note that the name of the component (hatch) now appears in the FeatureManager (Figure 6.27). The designation (f) means that the component is "fixed"; it is fully constrained in the assembly window, and cannot be moved or rotated. Note the Mates group at the bottom of the FeatureManager. As we define the geometric relations between components, these relations will be stored under this Mates group. The hinge component will now be brought into the assembly.

Note that the CommandManager has an Assemblies group containing many of the tools we will use in assembling the components.

FIGURE 6.28

Select the Insert Component Tool from the Assemblies group of the CommandManager. Select Browse, and find the hinge file. Click Open, and move the cursor to the approximate position shown in Figure 6.28. Click to place the hinge.

This will import the hinge component, with its origin located at the point selected. The exact position and orientation are not important at this point, since the Mate Tool will be used to establish the position and orientation with respect to the base part.

FIGURE 6.29

Note that the name of the hinge component now appears in the FeatureManager, with the (−) designation preceding it (Figure 6.29). This designation indicates that the component is "floating," and can be moved or rotated (as its degrees of freedom allow).

FIGURE 6.30

Although the CommandManager contains Move Component and Rotate Component Tools, versions of the software beginning with the 2004 release allow parts to be moved or rotated directly with click-and-drag operation.

Click on the hinge with the left mouse button, and, while holding the key down, drag the hinge to a new position, as shown in Figure 6.30.

Click on the hinge with the right mouse button and, while holding the key down, rotate the hinge to a new orientation, as shown in Figure 6.31.

More control over the move and rotate commands is available with the Triad Tool. The Triad Tool can be activated from the right-click menu of a component.

FIGURE 6.31

Right-click on the hinge (shown in Figure 6.32 in wireframe mode to better show the Triad Tool), and select Move with Triad from the menu.

The Triad Tool is shown in Figure 6.33. The arrows correspond to the principal axes of the assembly. If you click and drag with the left mouse button on one of the arrows, then the component moves in that direction. If you click and drag with the right mouse button on one of the arrows, then the component rotates about that axis.

Experiment with moving and rotating the hinge component with the Triad Tool. Place the hinge in the approximate position and orientation shown in Figure 6.34. Click in the white space around the hinge to turn off the Triad Tool.

FIGURE 6.32

FIGURE 6.33

While it is not strictly necessary to place the component in its approximate position and orientation prior to defining mate instructions, doing so can remove ambiguity and simplify the establishment of the mates.

The Mate Tool will now be used to establish the first geometric relationship between the hinge and the hatch.

FIGURE 6.34

FIGURE 6.35

Click the Mate Tool from the Assemblies group of the CommandManager.

This brings up the Mate PropertyManager.

Using the Rotate View Tool, rotate the view so that the bottom face of the hinge can be seen. Press the esc key to turn off the Rotate View Tool. Select the bottom face (Figure 6.35).

The name of the surface will appear in the pink area of the Mate dialog box.

If you select something incorrectly during a mate operation, you can clear the selection box at any time by right-clicking in the assembly window and selecting the Clear Selections option.

Return to a Trimetric or Isometric View, and select the top face of the hatch component (Figure 6.36).

FIGURE 6.36

FIGURE 6.37

In the Mate dialog box, the selected faces are shown, and a list of possible mates is shown (Figure 6.37). By default, a coincident mate is selected when two flat surfaces are selected. This means that the two selected faces will be coplanar. If the "Show preview" box is checked, then the hinge has moved to the position where it will be placed when the mate is accepted. If the "Show popup dialog" box is checked, then the small toolbar shown in Figure 6.36 appears in the modeling window.

The Mate Alignment Tools in the dialog box are important because there is often more than one configuration that meets the specification of the selected mate. For example, the hinge could be upside-down and the selected mate could still be satisfied. An advantage of placing and orienting a component before applying mates is that the default alignments of the mates are usually correct. However, we will illustrate the use of the Mate Alignment Tools before applying the mate to the hatch and hinge.

Select the Aligned Tool, as shown in Figure 6.38.

FIGURE 6.38

FIGURE 6.39

Note the position of the hinge, as shown in Figure 6.39. The mate conditions are still satisfied, but not in the manner desired.

Select the Anti-Aligned Tool, as shown in Figure 6.40.

To better view the effect of the mate, switch to the Right View.

FIGURE 6.40

As shown in Figure 6.41, the bottom of the hinge and the top of the hatch are coplanar.

Click the check mark in the Mate dialog box or the pop-up box to apply the mate.

Click and drag the hinge to the approximate position shown in Figure 6.42.

FIGURE 6.41

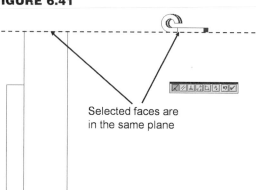

Selected faces are in the same plane

FIGURE 6.42

Note that the hinge component can be moved with respect to the fixed hatch component, but that the two mated faces remain coplanar. That is the geometric effect of the mate.

A second mate will now be added to provide additional location information for the hinge relative to the fixed hatch component.

By rotating the view and zooming in on the hinge, locate the assembly in the approximate position shown in Figure 6.43.

FIGURE 6.43

FIGURE 6.44

With the Mate dialog box still open (select the Mate Tool if you closed it accidentally), select the two faces shown in Figure 6.44. Click the check mark to apply this second coincident mate.

Switch to the Isometric View, and select the two faces shown in Figure 6.45.

In this case, we do not want to apply the default coincident mate. Rather, we want these faces to be a specific distance apart.

FIGURE 6.45

Select a distance mate from the Mate dialog box or the pop-up box, as shown in Figure 6.46. Set the distance to 1 inch.

FIGURE 6.46

A distance mate has two possible configurations. If you check the Flip Dimension box, then the other configuration is selected, as shown in Figure 6.47.

With the mate defined correctly, as in Figure 6.46, click the check mark to apply the mate.

FIGURE 6.47

Click esc to close the Mate dialog box. Try to move the hinge by clicking and dragging it.

FIGURE 6.48

A message appears, as shown in Figure 6.48, stating that the part cannot be moved. The three mates that we have applied have completely defined its position and orientation relative to the fixed hatch.

A second hinge will now be added to the assembly. Since the second hinge will be identical to the first, no new component must be created. A second instance of the first hinge can simply be added to the assembly.

FIGURE 6.49

Select the Insert Component Tool, and browse to the hinge file. Place the hinge into the assembly window, as shown in Figure 6.49.

Using a procedure similar to the one outlined previously in this chapter, add two coincident mates and a distance mate to fully constrain the second hinge in the position shown in Figure 6.50.

FIGURE 6.50

Once defined, mates can be easily modified. For instance, assume that a redesign specified that the location of the second hinge should be 4 inches from the mated edge rather than 1 inch.

Click the + sign next to the Mates entry in the FeatureManager. Locate the distance mate associated with the second hinge (called Distance2), and right-click on the mate name, as shown in Figure 6.51. Select the Edit Feature option.

The PropertyManager associated with this mate will reopen, and allow for editing of the mate.

FIGURE 6.51

FIGURE 6.52

Change the value of the distance to 4.00 inches, as shown in Figure 6.52. Accept the change by clicking the check mark.

The resulting position of the second hinge is shown in Figure 6.53.

In this way, a parameter of the mate can be readily redefined. Since we will want to continue with the 1 inch value rather than the modified 4 inch value, we will revert to the previous value.

Press the esc key to close the Mate dialog box. Click the Undo Tool to revert to the 1 inch dimension. Save the assembly file using the name "door," and close the file.

Note that the new file has the extension .SLDASM, indicating that it is a SolidWorks assembly file.

FIGURE 6.53

6.3 Creating a Complex Assembly of Subassemblies and Parts

In this section, we will create a new assembly using the door assembly created in the previous section as a subassembly. The assembly will be a model of a hinged hatch, shown in Figure 6.54.

The assembly will also make use of the hinge pin and frame parts created in Section 6.2.

FIGURE 6.54

Open a new assembly. Browse to the frame file, and place it at the origin (Figure 6.55).

In the next step, the door assembly created in the previous section will be used as a subassembly within the new assembly. An assembly can be inserted just like a part.

FIGURE 6.55

Select the Insert Component Tool, and click the Browse button. In the dialog box, change the "Files of type" option to assembly files, and locate the file door.SLDASM. Select it, and click Open (Figure 6.56).

Drop the subassembly into the assembly window in the approximate position shown in Figure 6.57 by clicking.

FIGURE 6.56

FIGURE 6.57

FIGURE 6.59

While the door is an assembly and not a part, it can be handled (mated, rotated, moved, etc.) just like a part component in the assembly window.

Right-click and drag on the door subassembly to rotate it into the approximate orientation shown in Figure 6.58, with the curved portion of the hinges toward the frame.

Select the Mate Tool, and select the side faces of both the hatch part of the door assembly and the frame, as shown in Figure 6.59. Define a 4.5-inch distance mate between the faces. Press the esc key to close the Mate dialog box.

Instances of the hinge parts will now be added to the frame.

Select the Insert Component Tool. Choose Browse, and change the "File of type" back to part files. Select the hinge part. Drop it into the assembly, and rotate and move the hinge to place it into the approx-imate location and orientation shown in Figure 6.60.

FIGURE 6.58

FIGURE 6.60

Click on the Mate Tool to begin a mate. Use the Rotate View Tool to reveal the back face of the hinge, and select it (Figure 6.61).

Return to the Trimetric View, and select the front face of the frame (Figure 6.62). Click the check mark to apply a coincident mate.

The next mate will align the faces of the two hinges to be engaged.

FIGURE 6.61

FIGURE 6.62

FIGURE 6.63

Select the face shown in Figure 6.63 on the hinge that was just mated to the frame.

FIGURE 6.64

Also select the corresponding face on the hinge mounted to the hatch, as shown in Figure 6.64. Click the check mark to apply a coincident mate.

FIGURE 6.65

The effect of the mate can be most easily seen from the Front View, as shown in Figure 6.65.

These two faces now lie in the same plane, but additional mates are required to fully constrain the desired relationship between these components. The hinges will now be brought together to share a common axis of rotation.

FIGURE 6.66

Select the cylindrical inner faces of each of the two hinges to be mated (Figure 6.66 and Figure 6.67).

FIGURE 6.67

By default, a concentric mate will be previewed between the two cylindrical faces, as shown in Figure 6.68.

Click the check mark to apply the mate.

The final mate required will locate the hinge on the frame.

FIGURE 6.68

Click and drag the hatch downward until the top of the hinge is below the top of the frame, as shown in Figure 6.69.

Select the top surfaces of the hinge and frame, and apply a 2-inch distance mate, as shown in Figure 6.70. Close the Mate PropertyManager.

FIGURE 6.69

The hinge mate is now fully defined. We can see the effect of the mates by attempting to move the hatch.

Click and drag on the hatch, as shown in Figure 6.71.

Note that the hatch has a degree of freedom about the hinge, but is constrained against all other types of motion.

FIGURE 6.70

FIGURE 6.71

Note also that **FIGURE 6.72**
the hatch can appar-
ently rotate through
the frame, as shown
in Figure 6.72. This
is due to the fact that
only geometric rela-
tions between the
entities have been
defined, but no true
physical characteris-
tics have been im-
parted to the objects.

(We will learn how to use interference detection to
limit the motion of the model in Chapter 7.)

For completeness, the hinge pin will now be added to the assembly.

FIGURE 6.73

**Select the Insert Component Tool and insert the pin
component into the assembly, as shown in Figure 6.73.**

To aid in the selection of small details, a *filter* is sometimes used. When a filter is active, only certain entities can
be selected. We are adding mates between faces, so a filter
that allows only faces to be selected will be helpful.

FIGURE 6.74

**From the Standard toolbar, select
the Toggle Selection Filter Toolbar Tool, as shown in Figure 6.74.**

FIGURE 6.75

**In the Selection Filter
toolbar, select the
Filter Faces Tool, as
shown in Figure 6.75.**

Whenever a filter is active, a filter icon
appears beside the cursor, as shown in
Figure 6.76.

FIGURE 6.76

**Select the Mate Tool, and select the cylindrical face of the pin, as shown in Figure 6.77.
(Note that with the filter set, you cannot select an edge.)**

FIGURE 6.77

**Select one of the cylindrical faces of the hinge, as
shown in Figure 6.78. (Either the outer or the
inner face may be selected; since they are concentric, the resulting mate will be the same.)**

FIGURE 6.78

**The pin and the hinge will align concentrically in the preview of the mate. If
necessary, change the alignment (from anti-aligned to aligned, or vice versa)
of the mate to orient the head of the pin appropriately, and drag the pin to
the position shown in Figure 6.79. Click the check mark to apply the mate.**

FIGURE 6.79

FIGURE 6.80

FIGURE 6.81

Select the face on the hinge shown in Figure 6.80, and then the underside of the head of the pin, as shown in Figure 6.81.

FIGURE 6.82

Click on the check mark to apply the mate, as shown in Figure 6.82. Close the Mate Property-Manager.

FIGURE 6.83

Using this procedure, duplicate the assembly of the second set of hinges. Note that the final distance mate used in the assembly of the first set of hinges will not be required, as the concentric mate between the hinges will both align the hinges axially and locate the hinge on the frame.

The assembly is now complete, as shown in Figure 6.83. It will be used in subsequent chapters, so it must be saved. However, it is a good idea to clear any filters when they are no longer needed.

Select the Clear All Filters Tool, as shown in Figure 6.84. Click on the Toggle Selection Filter Toolbar Tool, as shown in Figure 6.85, to close the toolbar.

FIGURE 6.84

FIGURE 6.85

Save the assembly in a file entitled "hatch assembly."

Problems

P6.1   Create a 6 inch × 2 inch × 2 inch block. Using the Hole Wizard, add a hole counterbored for a 3/8 inch socket head cap screw with a depth of 1.5 inches (see Figure P6.1a)0. Use a linear pattern to create a pattern of five evenly spaced holes in the block. The resulting block is shown in Figure P6.1b, and a section view is shown in Figure P6.1c.

FIGURE P6.1a

FIGURE P6.1b

FIGURE P6.1c

P6.2   Using the parts created in Chapter 6, create a working assembly model of a single hinge (Figure P6.2).

FIGURE P6.2

P6.3   Create solid models of a 5 ft. long, 1.5 in. diameter pole (Figure P6.3a), and a 4 in. diameter sphere with a 1.5 in. diameter hole in it (Figure P6.3b and Figure P6.3c). Using these models and the flange model created in Chapter 1 (Figure P6.3d), create an assembly model of a flagpole (Figure P6.3e).

FIGURE P6.3a

FIGURE P6.3b

FIGURE P6.3c

Section A-A

FIGURE P6.3d   **FIGURE P6.3e**

P6.4   Create solid models of the shaft segment shown in Figure P6.4a and Figure P6.4b, and the key shown in Figure P6.4c. Create an assembly with these two parts and the pulley model described in Problem P1.3. The finished assembly is shown in exploded state in Figure P6.4d and in normal state in Figure P6.4e. (Note: We will learn how to create an exploded view in Chapter 7.)

FIGURE P6.4a   **FIGURE P6.4b**

FIGURE P6.4c

FIGURE P6.4d   **FIGURE P6.4e**

P6.5   Create models of the 2×4 members shown in Figure P6.5a. (Note that 2×4s have actual finished dimensions of 1.5 × 3.5 inches.) Create an assembly from these 2×4s as shown in Figure P6.5b. Consider creating one or more subassemblies to reduce the total number of mates required.

FIGURE P6.5a

FIGURE P6.5b

P6.6   Using simple components (flat boards, 2×4s, etc.), design a shelving unit for your room or apartment. Customize your design so that some items that you own (TV, stereo, etc.) will fit on the shelves.

Advanced Assembly Operations

Introduction

In Chapter 6, the basic operations required to create a SolidWorks assembly were introduced. In this chapter, some advanced modeling, visualization, and analysis operations will be developed. The hatch subassembly and assembly models created in Chapter 6 will be used in the tutorials.

In this chapter, you will:

- add assembly-level features to an assembly model,
- use the Smart Fasteners Tool to add standard fasteners to the assembly model,
- create an exploded view of an assembly model, and
- use the interference detection and collision detection features to analyze the assembly model.

7.1 Adding Features at the Assembly Level

FIGURE 7.1

In Chapter 6, assemblies were created from pre-existing part files. It is sometimes desirable to make modifications to the parts in the assemblies. In this section, the addition of features at the assembly level will be described.

Open the assembly file entitled door.SLDASM created in Chapter 6, which is shown in Figure 7.1.

After switching to a top view, zoom in on the hole pattern of one of the hinges (Figure 7.2).

FIGURE 7.2

In this exercise, a hole pattern matching the hinge pattern will be created in the hatch.

Select the Assemblies group of the CommandManager. Click on the Features Tool, which reveals a menu of assembly-level features, as shown in Figure 7.3. Select the Hole Wizard.

FIGURE 7.3

A #10 hole (matching one of the holes in the hinge) will be made in the hatch.

Select the Hole tab, and create a 1 inch deep blind #10 hole (Figure 7.4).

Click the Next button, and a prompt for the hole location will appear.

FIGURE 7.4

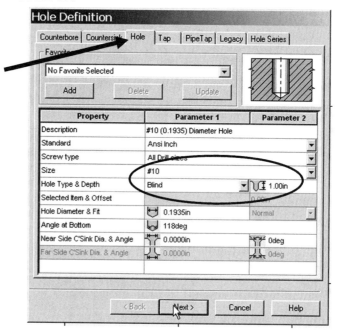

The hole will be temporarily located somewhere on the face of the hatch, then the location established using a concentric relation with one of the holes in the hinge.

Click somewhere on the face of the hatch (Figure 7.5).

FIGURE 7.5

The face of the hatch rather than the face of the hinge was selected, since the hinge already has holes. The depth of the hole will be set based on the selected surface. This will temporarily locate the hole, but a relation will be added to make this hole concentric with one of the hinge holes.

DESIGN INTENT

Since the holes that we are adding in the tutorial affect only the hatch part, we could have added them to the part file. Instead, we have added them at the assembly level, and the holes do not appear in the part model (verify this by opening the hatch file after adding one or more of the holes to the assembly). There are two types of features: *part-level* features, and *assembly-level* features. Where we apply the features should reflect the actual manufacturing and assembly processes. For example, if the holes are predrilled into the hatch before the assembly with the hinges, then the holes should be added to the part model and are considered part-level features. If the holes are added after placing the hinge on the hatch and using the

hinge's holes to locate the drilled holes in the hatch, then these are assembly-level features.

The difference between part-level and assembly-level features is especially important when creating detailed drawings. In this example, the part drawings would contain all information needed to manufacture and/or inspect the hinge and hatch parts. The assembly drawing would show which parts make up the assembly (in a *Bill of Materials*), and would contain only the dimensions necessary to assemble them. In this case, the dimensions to locate the hinges and the hole definitions would be defined on the assembly drawing (we will learn how to make an assembly drawing in Chapter 8).

FIGURE 7.6

FIGURE 7.7

Hit the esc key to close the PropertyManager.

The Hole Placement dialog box will remain open.

Select the point shown at the hole center you just created, and with the ctrl key depressed select the edge of the upper-left hole in the hinge (Figure 7.6).

In the PropertyManager, establish a concentric relation between the selected items. The hole will be previewed in the correct location (Figure 7.7).

Click the Finish button in the Hole Placement dialog box, and the hole will be created.

The hole is best viewed from the front view in a wireframe mode, as shown in Figure 7.8.

We will construct three more holes for this hinge by defining an Assembly Feature Linear Pattern. Since we want it to match the dimensions of the hole pattern in the hinge, we will need to recall the dimensions used to define this pattern.

FIGURE 7.8

FIGURE 7.9

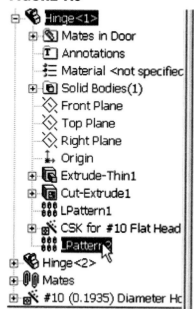

Go to a top view. In the FeatureManager, expand the first instance of the hinge component by clicking on the + sign next to the Hinge<1> entry.

The features associated with the hinge component will be shown, as in Figure 7.9.

Locate the entry LPattern2 (which follows the first hole definition), and double-click on it.

The dimensions associated with the linear pattern will appear, as in Figure 7.10. Note that there is a horizontal repeat of 2 inches, and a vertical repeat of 0.9 inch.

FIGURE 7.10

Press the esc key to deselect the hinge hole pattern.

Select the Features Tool from the Assembly group of the CommandManager, and select Linear Pattern from the menu that appears, as shown in Figure 7.11.

FIGURE 7.11

The LPattern PropertyManager will appear, with the Direction1 box highlighted in pink.

Select the top horizontal edge as Direction 1, set the distance to 2 inches, and the number of instances to be 2.

The Distance2 box will now be highlighted in pink.

Select a vertical edge as Direction2, set the distance to 0.9 inch, and the number of instances to be 2.

FIGURE 7.12

Direction Arrows

Click in the Features to Pattern box to highlight it in pink. Click on the #10 hole in the FeatureManager (which has "flown out" to the right of the PropertyManager) to select it as the Feature to Pattern. Change the pattern directions as necessary, so that the direction arrows are correct, as shown in Figure 7.12

Click the check mark in the PropertyManager to apply the linear pattern.

A hole pattern matching the holes in the hinge has now been created, as shown in Figure 7.13.

FIGURE 7.13

FIGURE 7.14

Repeat the entire operation on the other hinge.

The result is shown in Figure 7.14.

Save the changes to the assembly by selecting File: Save.

7.2 Adding Fasteners Using the Smart Fasteners Tool

The SolidWorks software has an add-in feature that can be used to intelligently insert appropriate mechanical fasteners into an assembly. It is possible that some limited-license SolidWorks products may not have this feature available. If this add-in is not available, this section may be skipped and Section 7.3 can be completed with very little alteration.

FIGURE 7.16

Be sure that the Toolbox and Toolbox Browser Add-Ins are available by selecting Tools: Add-Ins from the main menu. Scroll down to find the SolidWorks Toolbox and SolidWorks Toolbox Browser boxes (Figure 7.15). If they are not already checked, check them. Click OK to close this box.

FIGURE 7.15

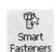

Smart
Fasteners

Select the Smart Fasteners Tool from the Assembly group of the Command-Manager.

In the PropertyManager, the Selection box will be high-lighted in pink. We will now specify the holes into which the fasteners will be inserted.

In the FeatureManager, expand the hinge parts and select the holes that were added with the Hole Wizard, as shown in Figure 7.16. Click Add in the PropertyManager.

FIGURE 7.17

After the selected holes are analyzed, fasteners will appear, as shown in Figure 7.17.

While the analysis performed by this intelligent design tool attempts to insert the appropriate fasteners for the design, the engineer has the option of overriding the choice and making modifications to the fastener selection. In our case, the fasteners chosen have a length consistent with the thickness of the hinges rather than with the total depth of the holes.

FIGURE 7.18

In the PropertyManager, the added fasteners are shown (Figure 7.18). Right-click on the first fastener group and select Properties.

Change the fastener length to 1 inch by selecting the correct length from the drop-down menu of standard lengths (Figure 7.19).

You can also change the appearance of the fasteners by selecting cosmetic threads from the options listed under Thread Display, if desired.

Click OK to accept the length change.

In the PropertyManager, a message appears that the fastener size does not match the hole, as shown in Figure 7.20. This is because the hole analyzed is in the hinge only, and the depth of the hole added into the hatch was not considered. This message can be ignored.

FIGURE 7.19

FIGURE 7.20   **FIGURE 7.21**

Change the length of the second fastener group.

Figure 7.21 shows the fasteners inserted into the holes in wireframe mode.

Save the changes to the assembly using File: Save, clicking Yes to rebuild before saving.

DESIGN INTENT

We saw earlier that features can be added at the part level or the assembly level. In the case of the holes into which our Smart Fasteners are added, we have combined a part-level feature (the holes in the hinges) with assembly-level features (the holes in the hatch). Therefore, the Smart Fastener Tool could not automatically select the proper length of fastener.

Often, holes at the assembly level are added to multiple components at the same time. For example, if we had hinges without holes, then they could be placed on the hatch and the holes *match-drilled* into both components. This is easily modeled in SolidWorks, as assembly-level operations can affect more than one component. If the holes were modeled this way, then the Smart Fastener Tool would select fasteners of the proper length.

Of course, this method would not be consistent with the way in which the hatch is to be manufactured. Keep in mind when deciding whether to add features to parts or assemblies that your model should represent the actual manufacturing steps.

7.3 Creating an Exploded View

Exploded views are often used to visualize assemblies. In this section, an exploded view of the door assembly will be created. Section 7.3.1 will demonstrate the first step in the exploded view, where the fasteners are exploded. Section 7.3.2 will show the explosion of the hinges. If you do not have the Smart Fasteners Tool and skipped Section 7.2, you should skip 7.3.1 and proceed with 7.3.2.

In either case, begin with a trimetric view of the door assembly (Figure 7.22).

FIGURE 7.22

FIGURE 7.23

Select the Exploded View Tool from the Assembly group of the CommandManager.

This will bring up the Assembly Exploder dialog box.

To initiate a new step in the exploded view, click the New Step button (Figure 7.23).

7.3.1 Exploding the Fasteners

If you have inserted the fasteners using the Smart Fasteners Tool, proceed with this section. If not, skip to Section 7.3.2. The dialog box will prompt for the "direction to explode along" by highlighting the box in pink.

FIGURE 7.24

Select the top face of the hatch component, as shown in Figure 7.24.

The direction of the explosion step will be normal to this surface, or upward.

The dialog box will now prompt for selection of the "components to explode." The fasteners can be selected either by clicking on them in the model window or by selecting them from the FeatureManager. An advantage of selecting patterned features from the FeatureManager is that only the first feature of the pattern needs to be selected; the other features in the pattern will be exploded in the same manner.

Expand the Smart Fastener entries in the FeatureManager by clicking on the + signs next to the entries. Select the Flat Head 100 Screw components by name (Figure 7.25).

FIGURE 7.25

Change the Distance entry in the dialog box to 4 inches (Figure 7.26), and click the check mark to apply the step.

The first step in the exploded view will appear (Figure 7.27), with the fasteners displaced upward by 4 inches.

Initiate the second step in the exploded view by again clicking on the New Step icon.

Proceed to the next section, where the hinges will also be exploded.

FIGURE 7.26

FIGURE 7.27

7.3.2 Exploding the Hinges

Note: If you did not add the fasteners using Smart Fasteners, and skipped Sections 7.2 and 7.3.1, the images on your screen will not include the fastener components shown in this tutorial. However, all the commands can be executed, and the images should match those on your screen otherwise.

The Assembly Exploder dialog box should be open on the screen. The dialog box will prompt for the "Direction to explode along" by highlighting the box in pink.

Select the top face of the hatch component as the direction to explode (Figure 7.28).

The "Components to explode" box will now be highlighted in pink, prompting for selection of the appropriate components.

FIGURE 7.28

FIGURE 7.29

FIGURE 7.30

Click on the hinges to select them. Change the explode distance to 2 inches, and click the check mark.

The results of this step are shown in Figure 7.29.

The explosion distance can also be modified directly by dragging the exploded components.

Click and hold on the green arrowhead associated with the current explosion step, as shown in Figure 7.30. With the mouse button still depressed, drag the exploded components to a new distance, then release the mouse button.

To revert to the previous 2 inch distance, click the Undo Tool in the Assembly Exploder dialog box, as shown in Figure 7.31.

FIGURE 7.31

FIGURE 7.32

Note that if you have defined multiple assembly steps, previous steps may also be modified at this point. The Previous Step and Next Step Tools (the arrows between the New Step and Undo Tools) can be used to view and modify the definitions of other assembly steps.

Click the OK button to implement the explosion step. Save the modified assembly.

The exploded view is shown in Figure 7.32.

Now that the exploded view has been defined, you may toggle between the exploded and collapsed views at any time.

In the FeatureManager, right-click on the name of the door assembly. Select Collapse (Figure 7.33).

The exploded view will be toggled to the collapsed view.

Right-click on the name again, and select Explode.

FIGURE 7.33

This will toggle the display back to the exploded state.
The assembly steps can still be modified at any time.

FIGURE 7.34

From the top of the current FeatureManager, select the Configuration-Manager tab (Figure 7.34).

The ConfigurationManager will appear.

Expand the entries in the ConfigurationManager so that the defined Explode Steps will appear (Figure 7.35).

In this exercise, there will be *two* steps if you performed the fastener explosion in Section 7.3.1, and *one* step if you skipped section 7.3.1.

FIGURE 7.35

Right click on one of the Explode Step entries, and select Edit Feature.

The Assembly Exploder dialog box will appear. The entries can be modified, and the Next Step and Previous Step Tools can be used to modify other explosion steps.

Experiment with changes if you desire. Undo any changes you made, and save the assembly, which will be used in the creation of an assembly drawing in Chapter 7.

Sketch lines can be added to the exploded view to show how the parts fit together. Although they are not really necessary for a simple assembly such as our door, they can be very helpful in more complex assemblies.

Select the Explode Line Sketch Tool from the Assembly group of the CommandManager.

A 3-D sketch will be opened.

Click on the cylindrical face of one of the fasteners near the bottom of the fastener, as shown in Figure 7.36. Then click on the cylindrical face of the corresponding hole in the hatch, as shown in Figure 7.37. Click the right mouse button (OK) to add a line between the selected items.

FIGURE 7.36   **FIGURE 7.37**

The first sketch line is shown in Figure 7.38.

Repeat for the other fasteners, as shown in Figure 7.39. Close the sketch.

FIGURE 7.38   **FIGURE 7.39**   **FIGURE 7.40**

The 3-D sketch containing the explode lines is stored in the ConfigurationManager with the associated exploded configuration, as shown in Figure 7.40.

Close the assembly window, without saving the last changes made.

FIGURE 7.41

FIGURE 7.45

Move

7.4 Detecting Interferences and Collisions

In the assembly mode, the SolidWorks program has the ability to check for interferences between components. It does this by determining locations where solid volumes overlap. This tutorial will demonstrate these capabilities.

FIGURE 7.42   **FIGURE 7.43**   **FIGURE 7.44**

Open the file "hatch assembly.SLDASM" created in Chapter 6. Click and drag on the hatch to rotate it into the approximate position shown in Figure 7.41. Press the esc key to deselect the hatch.

In doing this, we have intentionally introduced an interference between the door and the frame.

From the main menu, select Tools: Interference Detection.

The PropertyManager shows the parts for which interference is to be analyzed. By default, the entire assembly is selected.

Click the check mark to commence interference detection.

In the PropertyManager, the interference between the frame and hatch is identified, as shown in Figure 7.42. The interference is also highlighted in the model, as shown in Figure 7.43.

Click the × in the PropertyManager to end interference detection.

Although we have seen that components in an assembly can be moved by simply clicking and dragging them, the Move Component Tool allows collision detection to be incorporated into part movements.

Click and drag the hatch to the approximate position shown in Figure 7.44.

Select the Move Component Tool from the Assembly group of the CommandManager. In the PropertyManager, check Collision Detection and Stop at collision. Under Advanced Options, select Highlight faces and Sound, as shown in Figure 7.45.

Move
Compo...

Move the hatch toward the frame until it stops, as shown in Figure 7.46.

Note that when the hatch and frame faces touch, the faces are highlighted and a tone signifies a collision.

Now drag the hatch upward as far as possible, as shown in Figure 7.47.

FIGURE 7.46   **FIGURE 7.47**   **FIGURE 7.48**

The movement stops when the hinges collide, as shown in the right view of Figure 7.48.

Click the check mark or hit the esc key to deselect the Move Component Tool, and close the file.

Problems

P7.1   *For those who have the Smart Fastener Add-In.* Import the part model created in Problem P6.1 into a new assembly. Use the Smart Fasteners Add-In to add the appropriate socket head cap screws into the assembly.

P7.2   *For those who have the Smart Fastener Add-In.* Create an exploded view of the assembly created in P7.1.

P7.3   *For those who have the Smart Fastener Add-In.* Create an assembly from the frame part and two hinges, as shown in Figure P7.3a. The hinges should be located as shown in Figure P7.3b. Add #10 holes (1 inch deep) to the frame, aligned with the holes in the hinges, and insert screws to attach the hinges to the frame.

FIGURE P7.3a   **FIGURE P7.3b**

P7.4   Create a new hatch assembly, using the door assembly from Chapter 7, the frame/hinge assembly from P7.3, and two pins.

P7.5   Create an exploded view of the flagpole assembly model created in Problem P6.3

P7.6 Create an exploded view of the hinge model created in Problem P6.2.

P7.7 Add an angular dimension between the faces of the hinges in the hinge assembly model, as shown in Figure P7.7. Use the Move Component Tool with collision detection to move the hinge into its limiting positions. What is the total angle through which the hinge can be rotated?

(Answer: 284 degrees)

FIGURE P7.7

47.30°

P7.8 Create an exploded view of the pulley assembly model created in Problem P6.4.

P7.9 Open the flagpole assembly created in Problem P6.3. Modify the diameter of the pole component to 2.00 inches, and rebuild the assembly. Use the Interference Detection Tool to locate the interferences between the components in the rebuilt assembly.

Creating Assembly Drawings

Introduction

In Chapters 6 and 7, the development of solid models of complex assemblies from SolidWorks part files was described. In this chapter, the documentation of these assemblies through the use of 2-D assembly drawings will be introduced.

In this chapter, you will:

- create a 2-D assembly drawing,

- incorporate an exploded view into an assembly drawing,

- generate a Bill of Materials, and

- explore the associativity between part models, assembly models, and assembly drawings.

8.1 Creating an Assembly Drawing

In this section, an assembly drawing of the door assembly created in Chapter 6 and modified in Chapter 7 will be produced. In terms of generating the standard three-view drawing of the assembly, a procedure identical to that used to generate a drawing of a part could be used. However, a different procedure will be employed to demonstrate an alternate method of drawing creation.

Open the assembly file "door.SLDASM" (modified in Chapter 7). Make sure that the assembly is in the "collapsed" configuration. Select the Drawing icon from the Standard toolbar, as shown in Figure 8.1.

Click OK to accept the Drawing Template, as shown in Figure 8.2

In the Sheet Format/Size dialog box, uncheck the "Display sheet format" box, as shown in Figure 8.3.

FIGURE 8.1

The omission of a sheet format is done for simplicity. The use of sheet formats is described in Chapter 2, and a sheet format could certainly be applied here.

In the new drawing window, select the Standard 3 View Tool from the Drawings group of the CommandManager.

Drawings

Standard 3 View

175

FIGURE 8.2

FIGURE 8.3

FIGURE 8.4

Three standard drawing views will be displayed, as shown in Figure 8.4. (If necessary, turn off the display of origins by selecting View: Origins from the main menu.)

We will now add dimensions to the drawing. For an assembly drawing, we want to show only the dimensions associated with assembly-level features and operations.

Select the top view, as shown in Figure 8.5. From the main menu, select Insert: Model Items. Check the box labeled "Only Assembly," as shown in Figure 8.6. Click OK.

FIGURE 8.5

FIGURE 8.6

DESIGN INTENT

When importing dimensions into our assembly drawing, we selected the option labeled "Only Assembly," as opposed to "Entire Model." As a result, we imported dimensions associated only with assembly-level features. If the other option had been chosen, then the dimensions defining the components would have been imported as well. These dimensions could be edited, resulting in changes to the part files.

Including component dimensions in an assembly drawing is not usually recommended. One reason is that the components are defined in separate drawings, so adding the dimensions to the assembly drawings is redundant. Another reason is that a component may be used in multiple assemblies, so editing a component at the assembly level may produce unexpected changes to other assemblies.

The dimensions related to the hole patterns and to the placement of the hinges are imported, as shown in Figure 8.7.

Remember that the hole pattern dimensions relate to the holes in the *hatch*, not the holes in the hinges. Since these holes were placed to match up with those of the hinges, a note will be a better method to call out their positions. We will add such a note in the next section.

FIGURE 8.7

Delete all of the dimensions except for the 1-inch dimensions locating the hinges, as shown in Figure 8.8.

FIGURE 8.8

Although it is not necessary to show both dimensions, we will see later that we can adjust the locations of the hinges by changing these dimensions on the drawing. Since the two hinges were placed in the assembly separately, the two 1-inch dimensions are not related. Note that if we wanted to add an equation to the assembly model, then only one of the dimensions would be needed.

8.2 Adding an Exploded View

In Chapter 7, an exploded view of the door assembly was created. This exploded view can be easily added to the drawing.

Select the Model View Tool from the Drawings group of the CommandManager.

Click in any of the existing drawing views to designate the assembly, as shown in Figure 8.9.

FIGURE 8.9

FIGURE 8.10

FIGURE 8.11

FIGURE 8.12

The PropertyManager will now contain a list of available named views associated with the door assembly, as shown in Figure 8.10.

Select Trimetric from the list.

Click on the location in the drawing window where the exploded trimetric view will appear, as shown in Figure 8.11. If the program displays a message regarding isometric dimensions, click Yes in the dialog box.

With the new view selected, select More Properties from the PropertyManager (or right-click within the drawing view boundary).

The Drawing View Properties dialog box will appear.

Check the "Show in exploded state" box to change the currently selected view to an exploded view (Figure 8.12), and click OK.

FIGURE 8.13

FIGURE 8.14

FIGURE 8.15

The exploded view will now appear in the drawing window, as shown in Figure 8.13.

We will now add a note regarding the placement of the holes in the hatch.

Select the Note Tool from the Annotations group of the CommandManager.

Click on the edge of one of the holes in the exploded view. This will create a leader for the note, as shown in Figure 8.14.

Click on the approximate location of the note, as shown in Figure 8.15.

FIGURE 8.16

FIGURE 8.17

Drill # 10 holes 1.0 deep to match holes in hinges

Enter the text shown in Figure 8.16.

If desired, uncheck the Use Document's Font box in the PropertyManager and select a new font type and/or size.

Click in the drawing window to place the note, and then press the esc key to end the Note command. Click and drag the note to its final position, as shown in Figure 8.17.

The drawing is shown in Figure 8.18.

FIGURE 8.18

Drill # 10 holes 1.0 deep
to match holes in hinges

1.00   1.00

8.3 Creating a Bill of Materials

It is often desirable to generate a parts list, or Bill of Materials (BOM), associated with an assembly. The SolidWorks program can automatically create this list from an assembly file.

Select any of the drawing views. From the main menu, select Insert: Tables: Bill of Materials, as shown in Figure 8.19.

FIGURE 8.19

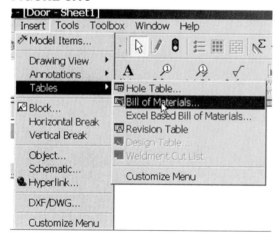

Accept the default selections listed in the PropertyManager, and click OK (Figure 8.20).

Click anywhere in the drawing to place the Bill of Materials, as shown in Figure 8.21.

The Bill of Materials, listing all parts in the assembly by name, along with the quantity and a description column (blank), will appear.

Right-click in the DESCRIPTION column, and select Delete: Column, as shown in Figure 8.22.

Click in any cell of the table. Select Table Format from the PropertyManager, as shown in Figure 8.23. Uncheck the "Use table's font" box and click the Font button. Edit the font type and size as desired.

FIGURE 8.20

FIGURE 8.21

FIGURE 8.22

FIGURE 8.23

FIGURE 8.24

FIGURE 8.25

Click and drag on column boundaries to adjust column widths, as shown in Figure 8.24.

Most companies use the part number, rather than a name, as the file name of a component or assembly. Therefore, the column displaying the file name is labeled "PART NUMBER." Since we have saved files with descriptive names rather than part numbers, we can change the heading in the Bill of Materials.

Click on the Part Number cell, and select Column Properties from the PropertyManager, as shown in Figure 8.25.

Edit the Title box to read PART NAME, as shown in Figure 8.26, and click the check mark.

Click and drag the title bar of the Bill of Materials to the position desired. Move the other drawing views as necessary, and add a title to the drawing as a note, as shown in Figure 8.27.

FIGURE 8.26

FIGURE 8.27

ITEM NO.	PART NAME	QTY.
1	Hatch	1
2	Hinge	2
3	CR–FHM1 0.19–24×1×1–N	8

Drill # 10 holes 1.0 deep to match holes in hinges

DOOR ASSEMBLY

FIGURE 8.28

Drill # 10 holes 1.0 deep to match holes in hinges

Our drawing is almost complete, but the item numbers in the Bill of Materials are not linked with the components in the drawing. We will add "balloons" with part numbers to the drawing.

Select the exploded Trimetric View. Select the Auto-Balloon Tool from the Annotations group of the CommandManager.

Balloons will be added to the view, as shown in Figure 8.28. The appearance of the balloons can be changed from the PropertyManager.

In the PropertyManager, select Tight Fit from the Size pull-down menu, as shown in Figure 8.29. Click the check mark to close the PropertyManager.

Click and drag the part numbers to relocate the balloons, as shown in Figure 8.30.

Save the drawing with the file name "door," and close it.

Note that the drawing file has the extension .SLDDRW.

FIGURE 8.29

FIGURE 8.30

Drill # 10 holes 1.0 deep
to match holes in hinges

8.4 Investigating Associativity

As with part drawings, assembly drawings maintain *associativity* with the assembly models used to create them; the drawing dimensions are linked to the model dimensions. Changes to assembly drawings will propagate to the assembly models; however, since only assembly-level dimensions are shown on assembly drawings, the part files will be unchanged. On the other hand, since assembly models are composed of part models, changes to part files will propagate to the assembly level and thus to the assembly drawing. In this section, this associativity will be demonstrated.

Open the door assembly file (door.SLDASM), the hatch part file (hatch.SLDPRT), and the door drawing file (door.SLDDRW). From the main menu, select Windows: Tile Horizontally to display all three windows at once, as shown in Figure 8.31.

FIGURE 8.31

In the drawing window, zoom in on the top view. Double-click on each of the 1-inch dimensions locating the hinges, and change the values to 3 inches. Click the Rebuild Tool, and the hinges will be displayed in the positions shown in Figure 8.32.

FIGURE 8.32

FIGURE 8.33

Click in the assembly window, and the assembly will be updated, as shown in Figure 8.33.

If you click in the part window, the hatch part will be unchanged.

We will change the hinge-location dimensions back to 1 inch from the assembly window.

In the assembly window, right-click on Annotations in the FeatureManager, as shown in Figure 8.34. Click on Show Feature Dimensions, and the dimensions will be displayed.

FIGURE 8.34

Double-click on each of the 3-inch dimensions and change them to 1 inch each. Rebuild the assembly, as shown in Figure 8.35.

FIGURE 8.35

If you click in the drawing window, you will see the drawing update to the new dimensions.

We will now see how a change to a part propagates to the assembly and the assembly drawing.

In the part window, double-click Extrude2 in the FeatureManager, as shown in Figure 8.36.

The dimensions associated with this feature (the raised portion of the hatch) will be displayed.

Double-click on the 1-inch dimension, and change it to 3 inches. Rebuild the model.

FIGURE 8.36

The updated model is shown in Figure 8.37.

Click in the assembly window. The message shown in Figure 8.38 will appear, letting you know that a component of the assembly has been changed. Click Yes to rebuild the assembly.

FIGURE 8.37

FIGURE 8.38

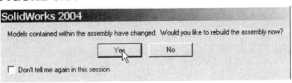

The result is shown in Figure 8.39.

Click in the drawing window.

The assembly drawing will show the modified component, as shown in Figure 8.40.

FIGURE 8.39

FIGURE 8.40

ITEM NO.	PART NAME	QTY.
1	Hatch	1
2	Hinge	2
3	CR–FHM1 0.19–24×1×1–N	8

Drill # 10 holes 1.0 deep to match holes in hinges

DOOR ASSEMBLY

Change the dimension back to 1 inch in the part window.

We will now change the name of the hatch component. Suppose that there are several other hatch sizes available, and so we want to give the 16 inch hatch a distinctive name.

In the part window, select File: Save As from the main menu. The message displayed in Figure 8.41 will be displayed.

This message is warning you that the part you are about to rename is used in assemblies. If you are creating a copy to use elsewhere, then you will want to use the "Save as

FIGURE 8.41

copy" option when saving the file. This will maintain the old file as the one that is used in the existing assemblies. If the "Save as copy" box is left unchecked, then the newly created file (with the new file name) will replace the old file in the assemblies. We will use the latter option, since we want the new file name to be used in our assembly.

Click OK to close the message box. In the Save As box, save the file with the file name Hatch16, as shown in Figure 8.42. Leave the "Save as copy" box unchecked. Click Save.

FIGURE 8.42

FIGURE 8.43

In the assembly window, notice that the new file name is referenced in the FeatureManager (Figure 8.43).

Switch to the drawing window, and rebuild the drawing.

FIGURE 8.45

FIGURE 8.44

ITEM NO.	PART NAME	QTY.
1	Hinge	2
2	CR–FHM10.19–24×1×1–N	8
3	Hatch16	1

Note that the part name of the hatch has changed in the Bill of Materials (Figure 8.44).

The order of the components in the bill has also changed. The balloons automatically change to reflect the part number as shown in the bill. By default, the parts are numbered in the order in which they are assembled. Since the new Hatch16 file replaced the other hatch file, it is considered to be the last component added. The item numbers can be changed manually.

Click in any cell of the Bill of Materials. In the PropertyManager, select BOM Contents, as shown in Figure 8.45.

FIGURE 8.46

FIGURE 8.47

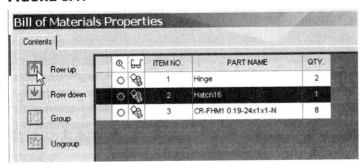

FIGURE 8.48

ITEM NO.	PART NAME	QTY.
1	Hatch16	1
2	Hinge	2
3	CR–FHM1 0.19–24×1×1–N	8

In the BOM Properties box, click in the empty box at the left end of the row corresponding to the hatch part. This will highlight that row, as shown in **Figure 8.46.**

Click the Row up button once to move the hatch up one row, as shown in **Figure 8.47.** Click Row up again to move the hatch into the top row. Click OK.

The BOM now shows the hatch as item #1 (Figure 8.48). Note that the balloons have automatically updated to reflect the new item numbers (Figure 8.49).

Close all of the windows. It is not necessary to save the changes.

FIGURE 8.49

Drill # 10 holes 1.0 deep
to match holes in hinges

Problems

P8.1   Create an assembly drawing, complete with exploded view and bill of materials, for the hinge assembly created in Problem P6.2.

P8.2   Create an assembly drawing, complete with exploded view and bill of materials, for the flagpole assembly created in Problem P6.3.

P8.3   Create an assembly drawing, complete with exploded view and bill of materials, for the shaft assembly created in Problem P6.4.

P8.4   Create an assembly drawing, complete with bill of materials, for the frame assembly created in Problem P6.5.

P8.5   *For those with the Smart Fasteners Add-In.* Create an assembly drawing, complete with exploded view and bill of materials, for the frame-and-hinge assembly created in Problem P7.3.

Using SolidWorks for the Generation of 2-D Layouts

Introduction

Earlier we learned to create 2-D drawings from solid parts and assemblies. For some tasks, however, working only in the 2-D environment is preferred. For example, floor plans and site drawings, plant equipment layouts, and electrical schematic drawings are usually created in 2-D. The SolidWorks program can be used for these applications, and the ease of changing dimensions allows multiple configurations to be quickly evaluated.

In this chapter, you will:

- choose and set the scale for a 2-D layout,
- use a unit system of feet and inches,
- create a floor plan and learn how to modify the layout to evaluate alternate configurations,
- expand your knowledge of using relations with 2-D entities,
- learn how to hide and show dimensions, and
- calculate the areas, centroid locations, and moments of inertias of 2-D areas.

9.1 A Simple Floor Plan Layout

In this exercise, we will prepare a layout drawing of a simple quality assurance lab for a manufacturing shop. The following items need to be placed in the lab:

- A tensile test machine, with a rectangular "footprint" of 4 feet wide by 3 feet deep
- An inspection bench, 8 feet by 3 feet
- A desk, 5 feet by 30 inches
- Three cabinets for storing measuring tools and fixtures, each 4 feet wide by 2 feet deep, with doors that swing outward

Parts will be brought in bins that are 4 feet square, so there will need to be room for one parts bin. The room that has been proposed for the lab is shown in Figure 9.1. We would like to prepare a layout drawing to show how the equipment should be placed in the room.

Begin by opening a new drawing in SolidWorks. When prompted to select a sheet format, pick A-Landscape as the paper size, and

FIGURE 9.1

make sure the "Display sheet format" box is unchecked, as shown in Figure 9.2. If the Model View dialog appears in the PropertyManager, clear the box labeled "Start command when creating new drawing," and click the × to end the Model View command.

When we import parts into drawings, an appropriate scale is automatically set. When preparing a 2-D layout, we must specify the scale. To make the room fill most of the paper, we can set the scale at 1 inch equals 2 feet. Thus, the 14-foot dimension will appear as 7 inches on the drawing.

FIGURE 9.2

FIGURE 9.3

Right-click anywhere in the drawing area, and select Properties. Set the scale to 1:24, as shown in Figure 9.3. Click OK.

We can use units of feet, inches, or mixed feet and inches for this layout. We will use mixed feet and inches.

FIGURE 9.4

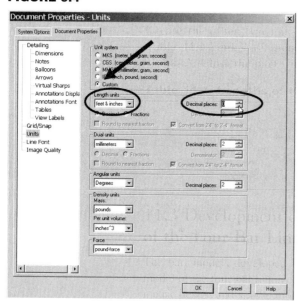

Select Tools: Options from the menu. Under the Document Properties tab, select Custom as the unit system, and feet & inches from the pull-down menu of length units. Set the decimal places to zero, as shown in Figure 9.4.

Select the Rectangle Tool from the Sketch group of the CommandManager, and drag out a rectangle. Dimension the sides of the rectangle, as shown in Figure 9.5. When entering the dimensions as feet, include the foot symbol (') after the number. Otherwise, the dimesion units default to inches.

With the Dimension Tool turned off, click and drag on one corner of the rectangle until the shape is centered on the sheet. Then select Fix in the PropertyManager to fix the point on the sheet, as shown in Figure 9.6.

FIGURE 9.5   **FIGURE 9.6**   **FIGURE 9.7**

Delete the bottom line and replace it with the two lines shown in Figure 9.7. Dimension each of the new lines, and add a collinear relation to the lines.   **FIGURE 9.8**

The drawing should be fully defined.

There are several ways that the doors and the arcs that represent their swing paths can be drawn and dimensioned. The method here uses relations to define the geometry.

Draw the vertical lines as shown in Figure 9.8.

Select the Centerpoint Arc Tool from the Sketch group of the CommandManager.

This tool allows you to construct an arc by picking the center of the arc and then its two endpoints.

FIGURE 9.9

Move the cursor directly over the intersection of one of the vertical lines and the adjacent horizontal line, as shown in Figure 9.9. Click and hold the left mouse button to drag out a circle to the other end of the vertical line, as shown in Figure 9.10. Release the mouse button.

FIGURE 9.10

Starting from the end of the line, click and drag the arc 90 degrees, as shown in Figure 9.11.

Check the "For construction" box in the PropertyManager.

Repeat for the other door, as shown in Figure 9.12.

FIGURE 9.11

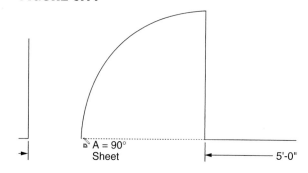

ᗅ A = 90°
Sheet

5'-0"

FIGURE 9.12

5'

FIGURE 9.13

Select both arcs, and add Tangent and Equal relations, which will join the arcs as shown in Figure 9.13.

The drawing should once again be fully defined.

In order to make the outer walls and doors stand out, we will make those lines thicker.

FIGURE 9.14

Line Thickness
Changes the thickness of
edges and sketch entities.

Show the Line Format toolbar by right-clicking in the toolbar area. The active toolbars are shown with checks. Click on Line Format. Select the lines representing the walls and doors. In the Line Format toolbar, select the Line Thickness Tool, as shown in Figure 9.14, and pick a thicker line than the default.

The walls and doors now stand out from the dimension and construction lines, as shown in Figure 9.15.

We will now place the tensile test machine in the room.

FIGURE 9.15

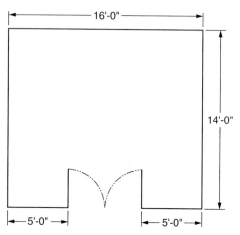

Draw a rectangle, and dimension it as 4 feet by 3 feet, as shown in Figure 9.16.

The dimensions that we added are not necessary to show in the drawing, but are required to set the size of the machine's footprint. We can hide the dimensions so they do not show, but still control the size of the rectangle.

Turn off the Smart Dimension Tool. Right-click on one of the dimensions, and select Hide, as shown in Figure 9.17. Repeat for the other dimension.

The rectangle is fully defined except for its position on the drawing. By clicking and dragging on one on the corners, you can move it.

FIGURE 9.16

3'-0"

4'-0"

FIGURE 9.17

Click and drag a corner of the rectangle until it is placed near the upper-right corner of the room, as shown in Figure 9.18.

FIGURE 9.18

Note: If you drag a corner until it contacts another point or a line, as shown in Figure 9.19, then a relation is automatically created with that entity, and you will be unable to move the rectangle away from the entity. If you desire to do so, then selecting the point will display existing relations in the PropertyManager, as shown in Figure 9.20. Select the Coincident mate and delete it.

FIGURE 9.19

FIGURE 9.20

While it is not necessary to fix the rectangle in place, doing so is a good idea in that it will make the drawing fully defined if all other required dimensions and relations are present. If you fix the position of the rectangle and the status bar reports that the drawing is still underdefined, then you will probably want to check existing dimensions and relations to see why.

Click on a corner of the rectangle to select that point and display its properties in the Property-Manager. Click Fix to fix that point.

The drawing should now be fully defined. We will now label the rectangle as the location of the tensile test machine.

FIGURE 9.21

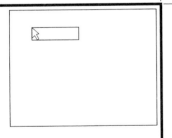

Select the Note Tool from the Annotations group of the CommandManager. Click at the approximate location of the note, as shown in Figure 9.21.

Type in the text of the note, as shown in Figure 9.22.

FIGURE 9.23

FIGURE 9.22

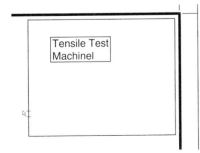

By default, the text will be left justified. If desired, center the text by choosing the Center Align button in the PropertyManager, as shown in Figure 9.23. Click anywhere outside of the note box, and then esc to end the note creation.

(If you want to place the same note in another location, then you can click the note down in multiple locations before using the esc key to end the process.)

FIGURE 9.24

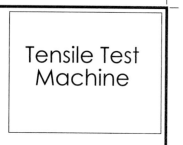

If you want to change the font or size of the note, then you can do that from the PropertyManager by unchecking the "Use document's font" box and selecting the Font button. However, if you want the note font changed for the entire drawing, do so from Tools: Options: Document Properties: Annotations Font: Note from the main menu.

FIGURE 9.25

Change the font type as size, if desired. Click and drag the note to its final position, as shown in Figure 9.24.

Add the first of the cabinets by drawing and dimensioning the rectangle as shown in Figure 9.25. Hide the dimensions.

After drawing the cabinet and moving its location around, you may decide that turning the cabinet 90 degrees will allow it to fit into the space better. The easiest way to do this is simply to switch the values of the dimensions. This will require showing and editing the currently hidden dimensions.

FIGURE 9.27

FIGURE 9.26

Select View: Hide/Show Annotations from the main menu. The hidden dimensions will be shown in gray. Click on each dimension that you want to show, as shown in Figure 9.26. Select Hide/Show Annotations again or esc to return to the editing mode. Change the cabinet dimensions as shown in Figure 9.27.

To hide the dimensions again, we can either right-click on each and select Hide, as we did earlier, or to hide several dimensions at once, we can use either Hide/Show Annotations from the menu or the Hide/Show Annotations Tool.

FIGURE 9.28

Select View: Hide/Show Annotations from the main menu. Click on each dimension to be hidden, as shown in Figure 9.28. Dimensions selected will be shown in gray. Press the esc key to return to the editing mode, and the selected dimensions will be hidden.

Add a note identifying the cabinet. If desired, rotate the text 90 degrees by changing the angle in the PropertyManager, as shown in Figure 9.29. Click OK. (A message appears in a pop-up box that tells you that the "Use document's font" box will be cleared. Click OK to close this message box.)

Add the other two cabinets. Dimension and label them, and fix their positions on the drawing, as shown in Figure 9.30.

FIGURE 9.29

FIGURE 9.30

FIGURE 9.31

Note that if you want to use Copy and Paste to add the cabinets, then the dimensions defining their sizes must also be copied, or else the cabinets will change size when dragged into a new position. Hidden dimensions cannot be copied, so you must first unhide the dimensions before using Copy and Paste commands.

Add and dimension rectangles representing the inspection bench and desk, as shown in Figure 9.31.

Label the inspection bench and desk. Add and label the 4-foot-square parts bin, with the lines representing it shown as construction lines. Fix all of the entities in position so that the drawing is fully defined. Hide the dimensions defining the bench, desk, and parts bin. The finished drawing is shown in Figure 9.32.

FIGURE 9.32

Finally, suppose that we wish to rotate the tensile test machine 45 degrees to give the operator better access to the machine. To do so, it will be necessary to delete some of the relations created automatically when the rectangle was drawn, and to add some new relations to maintain the rectangular shape and orient the rectangle properly.

Industrial Engineering

In this example, we created a floor plan to see if the proposed lab space would accommodate the equipment required for the lab. Consider the challenge of planning the layout of a manufacturing plant with hundreds of thousands of square feet of floor space and hundreds of machines. If not well planned, the operations of the plant will be crippled by inefficiencies in the ways that raw materials and parts move through the plant. Efficient plant layout is one of the functions of industrial engineers.

Industrial engineers perform many other functions toward the goal of improving operations. These functions might include monitoring and improving the quality of finished products, streamlining material handling and product flow, or redesigning work cells for better efficiency.

While the word "industrial" refers to the manufacturing environment where industrial engineers have traditionally worked, the skills of industrial engineers are now being widely used in service-sector businesses, as well. For example, many hospitals use industrial engineers to help improve quality and efficiency. Package-delivery companies, facing monumental logistics challenges associated with delivering packages worldwide under extreme time pressure, also use the services of industrial engineers.

Industrial engineers also are involved with product design, usually from the standpoint of *ergonomics*, the consideration of human characteristics and limitations in the design of products. In this area, there is overlap with the functions of the industrial designer.

Most engineering students in other disciplines take some coursework in industrial engineering topics. These topics could include engineering economy, quality control, project management, and ergonomic design.

FIGURE 9.33

FIGURE 9.34

FIGURE 9.35

Select the point of the tensile test machine rectangle that is currently fixed, as shown in Figure 9.33. (If you are not sure which point is fixed, click on each corner until a "Fixed" relation appears in the PropertyManager.) Highlight the "Fixed" relation in the PropertyManager, as shown in Figure 9.34, and delete it.

Click on each of the lines forming the rectangle, and delete the horizontal or vertical relation associated with each.

Select any two adjacent lines, as shown in Figure 9.35. Add a perpendicular relation. Repeat for two other pairs of lines.

FIGURE 9.36   **FIGURE 9.37**

The addition of these perpendicular relations defines the shape as a rectangle, but the removal of the horizontal and vertical relations allows it to be rotated. The hidden dimensions still apply to the rectangle, and fix its size.

Select the Smart Dimension Tool. Select one of the vertical lines of the rectangle and then a horizontal wall, as shown in Figure 9.36.

This will create an angular dimension between the lines, as shown in Figure 9.37. Change the dimension to 135 degrees, reflecting a rotation of 45 degrees, as shown in Figure 9.38.

FIGURE 9.38   **FIGURE 9.39**

Rotate the text labeling the machine −45 degrees, as shown in Figure 9.39.

Hide the angular dimension, and fix one of the corner points.

The drawing is now complete, as shown in Figure 9.40. Dimensions can be added to precisely place the objects in the room, if required. (It will be necessary to delete the fixed relations of the points if these dimensions are to be added.)

FIGURE 9.40

9.2 Finding the Properties of 2-D Shapes

The SolidWorks program can be used to determine the properties of areas. For example, the areas of complex shapes can be determined, and the locations of *centroids* and the values of *moments of inertia* of areas can be computed.

9.2.1 Calculating the Area of a Shape

FIGURE 9.41

Consider the building lot shown in Figure 9.41. Suppose that we want to find the acreage of this lot. The first step is to draw the lot. In order to fit the drawing on an 8-1/2 × 11 inch sheet, the minimum scale factor possible will be 1320 inches (110 feet) divided by 8.5 inches = 155. We will use a scale factor of 1:200.

Open a new drawing. Choose an A-Landscape paper size without a sheet format displayed. Set the drawing scale to 1:200 and draw and dimension the lot shown in Figure 9.41. Fix one corner of the lot, and the drawing should be fully defined.

Press the esc key to deselect any points or lines. Select Tools: Section Properties from the main menu. Click the Options tab, and set the units to feet and the number of decimal places to 6, as shown in Figure 9.42. Click OK.

FIGURE 9.42

FIGURE 9.43

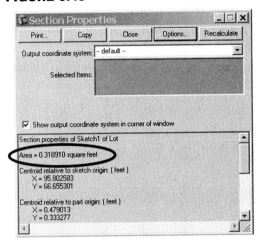

The area will be displayed in the units selected, in this case square feet. However, the area displayed will be that of the *drawing* area, not the actual area. Since the area of the drawing in square feet will be small, a large number of decimal places will be needed for accuracy.

The area displayed is 0.318910 square feet, as shown in Figure 9.43. To determine the area of the lot, the calculated area must be adjusted to account for the scale. Since area has the units of length squared, the calculated area must be multiplied by the scale factor squared:

$$\text{Area} = (0.318910 \text{ ft.}^2)(200^2) = 12{,}756 \text{ ft.}^2$$

To convert to acres, the conversion factor of 43,560 ft.²/acre must be applied:

$$\text{Area} = (12{,}756 \text{ ft.}^2)/(43{,}560 \text{ ft.}^2/\text{acre}) = 0.29 \text{ acres}$$

9.2.2 Calculating the Section Properties of a Shape

In mechanics of materials, it is often necessary to compute the moment of inertia of the cross-section of a structural member. For simple shapes, such as a rectangle or circle, easy formulas can be used for this computation. For compound shapes, such as a T-beam constructed from two rectangular shapes, the calculations are lengthier.

FIGURE 9.45

FIGURE 9.44

Open a new drawing. Choose A-Landscape paper without a sheet format displayed. Draw and dimension a rectangle, as shown in Figure 9.44.

Add and dimension the second rectangle shown in Figure 9.45, making sure to snap the first point of the new rectangle to the edge of the first rectangle.

Mechanics of Materials

The study of mechanics of materials involves determining *stresses, strains,* and *deformations* in bodies subjected to various loadings. Stress is the force per unit area acting on a point in a body, and the calculation of stress is necessary to predict failure of a structure. Strain is a measure of geometrical changes within a body. Stress and strain are related by properties of the material used. Deformations are related to strains in the body. While strain applies at a given point, deformation is the dimensional change of the entire body. For example, consider a 1 × 6 inch wooden plank resting on sawhorses near its ends. If the *span* of this beam is 6 feet, and a 175-pound man stands in the middle, how much will the plank move downward?

72 in.

175 lb.

Using mechanics of materials concepts, it can be determined that the deflection of such a beam is:

$$\delta = \frac{PL^3}{48EI}$$

where: P = force = 175 lb.

L = span = 72 in.

E = *modulus* of the material. A typical value for wood = 1,500,000 lb./in.²

I = moment of inertia of the cross-section

For a rectangular shape, the moment of inertia is:

$$I = \frac{1}{12} bh^3$$

So for our plank, with base b = 6 inches and height h = 1 inch, I = 0.50 in.⁴, and the deflection = 1.81 inches. If we could carefully place the plank so that it is resting on the 1-inch edge, then the dimensions of b and h are switched. The moment of inertia increases to 18 in.⁴ and the deflection reduces to 0.05 inches. Therefore, the *bending stiffness* of the plank has increased by a factor of 36 by changing its orientation. The moment of inertia is increased by moving as much material as possible as far away from the *centroidal axis* (the axis passing through the centroid, or center of area, of the section). With wood construction, this is done by aligning the members appropriately, as with floor beams that are placed with their long dimensions perpendicular to the floor. In steel construction, wide-flange beams maximize bending stiffness by placing most of the material in the flanges, as far from the centroidal axis as possible.

When complex shapes are used, calculation of the moment of inertia can be cumbersome. First, the centroid must be located, and then the moments of inertia of simple regions of the shape must be calculated and adjusted for their distances away from the centroidal axis. Finally, the moments of inertia of the individual regions are summed.

The Section Properties Tool can be a useful tool for calculating and/or checking the values of moment of inertia.

To center the second rectangle relative to the first, we could add a dimension. A better way is to add the centerlines shown in Figure 9.46, and add an Equal relation between them.

To compute the area or moments of inertia of the shape, it is not necessary to locate the drawing on the page. However, if we want to know the location of the centroidal axes, then we need to set a point at a known location.

Select the lower-left point of the second rectangle, as shown in Figure 9.47.

FIGURE 9.46

FIGURE 9.47

FIGURE 9.48

In the PropertyManager, set the coordinates of the point to x = 5 and y = 1. Fix the point, as shown in Figure 9.48.

In drawings, the lower-left corner of the page is the origin of the x-y coordinate system.

FIGURE 9.49

Select Tools: Section Properties from the main menu. A message will be displayed that the sketch has intersecting contours. Close the Section Properties dialog box.

The properties can be computed only for a sketch or drawing containing a single, closed contour.

Select the Trim Entities Tool from the Sketch group of the CommandManager. Trim away the overlapping portions of both rectangles, as shown in Figure 9.49.

Select Tools: Section Properties from the main menu.

The section properties are displayed, as shown in Figure 9.50.

There are two principal moments of inertia calculated. The moment labeled "L-xx" is the moment of inertia about an axis through the centroid of the cross-section, parallel to the X axis. The location of this axis is given by the Y coordinate of the centroid, 5.75 inches. Since a point at the bottom of the section was set at Y = 1 inch, the axis is 4.75 inches above the bottom of the section, as shown in Figure 9.51. The moment of inertia about this axis is 55.25 in.[4]. The other principal moment of inertia is about the axis shown in Figure 9.52. The value of the moment of inertia about this axis is 18.50 in.[4].

FIGURE 9.50

FIGURE 9.51

FIGURE 9.52

Problems

P9.1 Draw this 6-inch-tall five-pointed star using relations and hidden dimensions so that all sides are equal length and all angles are equal. Fix one point on the drawing; your drawing should be fully defined but not overdefined.

P9.2 Place the items from section 9.1 into the space shown here, which includes two 18-inch-square columns.

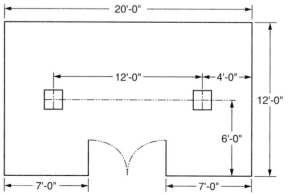

P9.3   Sketch a floor plan of your room or apartment, showing the locations of furniture.

P9.4   A builder desires to construct a house on the irregular lot shown here. The desired house will be rectangular in shape, 55 feet by 40 feet, with the front of the house 55 feet wide. The front of the house is to be parallel to the street. If local regulations require that there be 10 feet between property lines and any point on the house, can the proposed house be built on the lot?

Note: Three of the angular dimensions will be driven dimensions, since the length of all sides and one angle defines the lot shape. However, when property is surveyed, all lengths and angles are measured. Trigonometry is then used to ensure that the dimensions are consistent to within a certain tolerance. This method allows for incorrectly measured or recorded lengths and angles to be detected. The angles are measured in degrees and minutes (1 minute = 1/60 degree).

Hint: Use the offset entities tool to offset the property lines 10 feet inward. You can then check to see if the house plan fits within the offset lines.

P9.5   Find the number of square feet in the lot described in P9.4.

(Answer: 6506 ft.²)

P9.6   Find the location of the centroid and the principal moments of inertia of the channel shape shown here. Make a sketch showing the locations of the centroidal axes.

Application of SolidWorks to Vector Mechanics

Introduction

Many engineering problems involve the manipulation of *vectors*. A vector is a representation of a quantity that is defined by both a magnitude and a direction, as opposed to a scalar quantity that can be defined by a magnitude only. For example, the speed of an object is a scalar quantity, while velocity is defined by both the speed and the direction of the object's motion. The SolidWorks 2-D drawing environment allows for easy graphical solution of vector problems.

In this chapter, you will:

- learn how to add vector quantities graphically,
- work with driving and driven dimensions,
- learn to solve for any two unknowns (magnitudes and/or directions) in a vector equation, and
- perform position analysis for some common mechanisms.

10.1 Vector Addition

Consider the two forces acting on the hook as shown in Figure 10.1. We would like to find the resultant force, that is, a single force that affects the hook in an equivalent manner to the two forces. The resultant force is the vector sum of the two vectors **A** and **B**. (Vector quantities are usually denoted by bold type or by a bar or arrow over the symbol: $\mathbf{A}$ or $\overline{A}$ or $\vec{A}$.)

Analytically, the vectors can be added by adding the *components* of the two vectors. If we define an x-y coordinate system, then each vector can be broken into its x and y components, as shown in Figure 10.2.

FIGURE 10.1

FIGURE 10.2

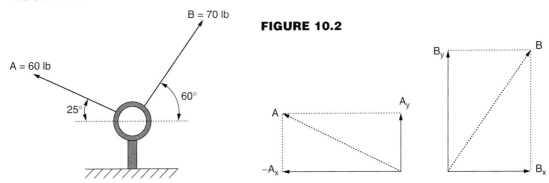

The components of the resultant vector **R** are found by adding the components of **A** and **B**:

	x component	y component
A	$-60\cos(25°) = -54.38$ lb.	$60\sin(25°) = 25.36$ lb.
B	$70\cos(60°) = 35.00$ lb.	$70\sin(60°) = 60.62$ lb.
R	-19.38 lb.	85.98 lb.

From the components, the magnitude and direction of **R** can be determined (see Figure 10.3):

$$R = \sqrt{(19.38)^2 + (85.98)^2} = 88.14 \text{ lb.}$$

$$\theta = \tan^{-1}\left(\frac{85.98}{19.38}\right) = 77.3 \text{ deg}$$

FIGURE 10.3

10.2 Vector Addition with SolidWorks

Open a new drawing. Choose A-Landscape as the paper size, with the "Display sheet format" box unchecked.

We must now choose scales: one for the magnitude of the vectors and another for the drawing. The vectors must be drawn to scale so that the length is proportional to its magnitude. If we choose to let 1 inch represent 10 pounds of force, then our two vectors will be 6 and 7 inches long. To fit onto our 8-1/2 × 11 inch drawing, it will probably be necessary to draw the vectors at one-half scale.

FIGURE 10.4

FIGURE 10.5

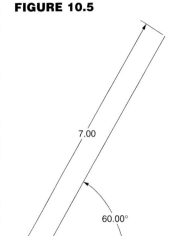

Right-click anywhere in the drawing area, and select Properties from the pop-up menu. Change the scale to 1:2.

(The default scale is defined in the template selected when creating the drawing.)

Select the line tool, and drag out a diagonal line to represent the first vector. From the starting point of the first vector, drag out a horizontal centerline, as shown in Figure 10.4.

Select the Smart Dimension Tool, and click once on the vector line to set the length to 7 inches. Click once on the vector line and once on the horizontal line to set the angle to 60 degrees, as shown in Figure 10.5.

FIGURE 10.6

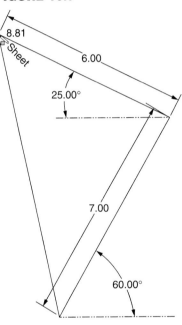

FIGURE 10.7

To add vectors graphically, we set them "tip-to-tail." That is, the second vector starts at the end of the first vector.

Draw the second vector line diagonally from the end of the first vector. (Be sure not to align this vector line with any of the dashed lines that appear. Doing so will add a constraint to the line that will have to be deleted before the vector's magnitude and direction can be defined.) Add another horizontal construction line, and add the 6-inch and 25-degree dimensions as shown in Figure 10.6.

The resultant vector extends from the starting point of the first vector to the ending point of the second vector.

Add the line representing the resultant vector, as shown in Figure 10.7, snapping to the endpoints.

Add another horizontal construction line, and add the dimensions for the length and orientation angle of the resultant vector. When you add these dimensions, you will get a message that adding this dimension will make the drawing over-dimensioned, and asking if you want to make the dimension driven or leave it as driving (Figure 10.8). Click OK to leave each dimension driven.

A driving dimension is one that helps control the size and position of the drawing or sketch entities, while a driven dimension does not.

FIGURE 10.8

FIGURE 10.9

FIGURE 10.10

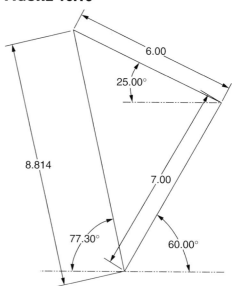

If the length of the resultant vector is not given to 3 decimal places, select that dimension and choose .123 from the pull-down menu in the PropertyManager, as shown in Figure 10.9.

The completed drawing is shown in Figure 10.10. The length of the resultant vector is 8.814 inches, corresponding to a magnitude of 88.14 pounds. This value and the angle of 77.3 degrees relative to the horizontal agree with the results found analytically.

10.3 Modifying the Vector Addition Drawing

The drawing just created is fine for calculating the magnitude and direction of the resultant vector, but visually is not clear. Vectors are usually shown with arrowheads to indicate their directions, and the line format of the vector lines can be changed to make them stand out from the construction and dimension lines. Also, the dimensions of the resultant vector can be shown differently to make them standout from the input dimensions. We will modify the drawing to make it more understandable, and to allow easy modifications of the input quantities.

For this exercise we will use the Font and Line Format toolbars.

FIGURE 10.11

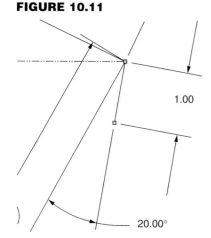

Right-click anywhere on any of the open toolbars. A list of available toolbars is displayed, with checks next to those already displayed. Click on both "Font" and "Line Format."

We will first add arrowheads to the vectors. We will use a simple arrowhead: a single line segment.

Add a line to the end of the 7-inch vector line. Add a length dimension of 1 inch and an angular dimension of 20 degrees, as shown in Figure 10.11.

FIGURE 10.12

FIGURE 10.14

Select both the vector line and the arrowhead line. From the Line Format toolbar, click on the Line Thickness Tool, as shown in Figure 10.12, and choose a thicker line style. If desired, change the color of the lines, using the Brush Tool to the left of the Line Thickness Tool.

These lines now stand out, as shown in Figure 10.13.

Hide the 1-inch and 20-degree dimensions, as shown in Figure 10.14.

Add and dimension arrowhead lines to the other vector lines. Hide the dimensions, and change the line thicknesses and colors, if desired, of the vector and arrowhead lines (Figure 10.15).

Use a different color for the resultant vector to make it stand out from the other vectors.

FIGURE 10.13

FIGURE 10.15

FIGURE 10.16

FIGURE 10.17

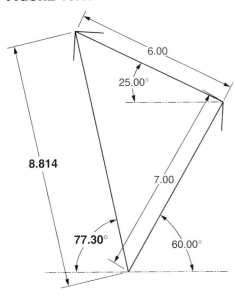

Modify the text of the dimensions by selecting them and changing the point sizes from the Font toolbar. Make the dimension text associated with the resultant vector larger than those of the other vectors. You can also make these text items bold by clicking the "B" Tool on the Font toolbar, as shown in Figure 10.16.

The completed vector drawing is shown in Figure 10.17.

One of the main benefits of using the SolidWorks program to add vectors is that the vector drawing can be easily modified to solve different problems. We will use our drawing to add a different pair of vectors.

FIGURE 10.18

To make this drawing more useable, define the orientation of the vectors from a common reference: the +x direction.

Delete the 25- and 70.7-degree dimensions and associated reference lines. Add new construction lines and dimensions as shown in Figure 10.18. Add the 155-degree dimension first so that the 102.7-degree dimension is the driven one.

Dimensions can be changed from driven to driving or from driving to driven by right-clicking on the dimension and checking or unchecking Driven. Of course, if too many dimensions are driving, the drawing will be overdefined.

Let's use this drawing to add the two vectors shown in Figure 10.19. We will let 1 inch equal 1 pound for the vector's scale.

Change the dimensions of the input vectors to 22 inches and 110 degrees (from the horizontal) for the first vector, and 45 inches and 15 degrees for the second vector.

Of course, the drawing now extends beyond the edges of the sheet.

FIGURE 10.19

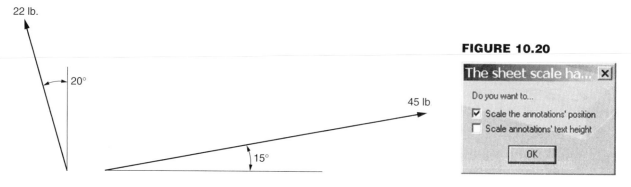

FIGURE 10.20

Right-click anywhere within the sheet borders, and select Properties. Set the scale to 1:5.

In the dialog box that appears, leave the check boxes as shown in Figure 10.20 and click OK.

This will allow the relative position of the dimension to stay the same, without changing the size of the text.

Drag the drawing into the sheet boundaries, and drag the dimensions into desired positions, as shown in Figure 10.21.

The resultant vector has a magnitude of 48.3 pounds, and is oriented at 42 degrees counterclockwise (CCW) from the x axis.

FIGURE 10.21

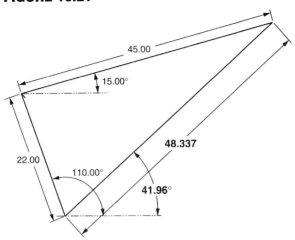

10.4 Further Solution of Vector Equations

In the previous example, two vectors were added to find the resultant vector's magnitude and direction. In a vector equation, *any* two unknowns (magnitudes and/or directions) can be determined. The following examples illustrate this concept.

A small plane can travel at an airspeed of 300 miles per hour. The flight path is to be at a heading of 15 degrees. (Heading is the angular direction measured CW from due North.) The wind is blowing from the WNW, as shown in Figure 10.22, at 60 mph. Find the plane's ground speed and the direction of the plane's travel relative to the air.

The vector equation for this problem is:

FIGURE 10.22

$$\vec{V}_{plane} = \vec{V}_{air} + \vec{V}_{plane/air}$$

where:

- $\vec{V}_{plane}$ = absolute velocity of the plane (ground speed)—magnitude unknown and direction known

- $\vec{V}_{air}$ = wind velocity—magnitude and directions known

- $\vec{V}_{plane/air}$ = velocity of the plane relative to the air (airspeed)—magnitude known and direction unknown

Open a new A-size drawing and set the drawing scale to 1:50.

We will let 1 inch equal 1 mile per hour.

FIGURE 10.23

Begin by drawing and dimensioning the vector representing the wind speed, as shown in Figure 10.23 near the bottom of the sheet.

It is helpful when we are dragging vectors together later to have a point fixed on the drawing.

FIGURE 10.24

FIGURE 10.25

Select the starting point of the first vector and click on the Fix icon in the PropertyManager.

If you need to move this point later, click on the point and the Fix relation will be listed in the PropertyManager. Select this relation and use the delete key to remove it. The point can then be dragged to a new location.

Add the airspeed vector of 300 mph at an arbitrary angle, as shown in Figure 10.24.

Add the resultant (ground speed) vector at 15 degrees from vertical. The length is unknown, but make the line long enough so that the airspeed vector will intersect it when rotated into position, as shown in Figure 10.25. If desired, change the color of the resultant vector line to make it easy to distinguish from the airspeed and wind vectors.

FIGURE 10.26

FIGURE 10.27

Click and drag the endpoint of the airspeed vector until it intersects with the ground speed vector (see Figure 10.26).

With the Trim Tool, trim away the end of the ground speed vector, as shown in Figure 10.27.

Add dimensions for the ground speed magnitude and the direction of the airspeed, as shown in Figure 10.28.

FIGURE 10.28

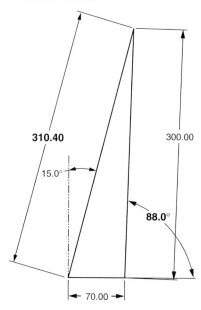

The result is that the ground speed is 279 miles per hour. Relative to the wind, the plane must fly about 4 degrees east of due north to achieve the desired course.

Variations of this problem can be easily solved by changing the input quantities. For example, consider the case where the wind is blowing from due west at 70 mph.

Change the dimensions defining the wind speed vector, as shown in Figure 10.29.

Figure 10.29 shows the resulting vector equation. The ground speed is now 310 miles per hour, as the wind contributes to the east-to-west component of the plane's travel.

FIGURE 10.29

10.5 Kinematic Sketch of a Simple Mechanism

In Chapter 11, we will look at the application of the SolidWorks program to model mechanisms, with assemblies of 3-D component parts. Often, the first step in the design of a mechanism is the preparation of a *kinematic sketch*, a 2-D drawing showing simplified representations of the members. For example, a four-bar linkage, which is a common mechanism used in many machines, can be represented by four lines, as shown in Figure 10.30.

In a *kinematic analysis*, the velocities and accelerations of the links and points are calculated, based on the velocity and acceleration of the driving link. For example, if Link 2 is connected to a motor and rotated about its pivot point at a constant velocity, the angular velocities and accelerations of the other two moving links and the translational velocities and accelerations of those links can be calculated. The accelerations are important because force is proportional to acceleration. If the accelerations are known, the forces acting on the members and joints can be calculated. (Note: Although there are only three moving members, this mechanism is referred to as a four-bar linkage because it is connected to a fixed or ground link, which is usually called Link 1.)

FIGURE 10.30

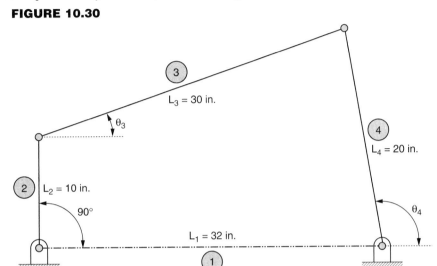

The first step in any kinematic analysis is a *position analysis*. For a given position of the driving link (the angular position of Link 2, 90° in the example), the positions of the other links must be calculated. Obviously, these positions can be calculated using trigonometry. For many mechanisms, a position analysis using trigonometry is surprisingly complex. A graphical solution is often utilized. The SolidWorks program is an excellent tool for graphical position analyses, in that dimensions can be easily changed and the effects on the rest of the linkage can be determined.

The links are often represented by vectors, as shown in Figure 10.31. In this case, all of the vector magnitudes (lengths) are known, as well as two of the vector angles (Link 1 is horizontal, and we will be performing the analysis for a given orientation of Link 2, the driving link). Therefore, we can solve the vector equation for the two unknown quantities, the angles θ_3 and θ_4.

FIGURE 10.31

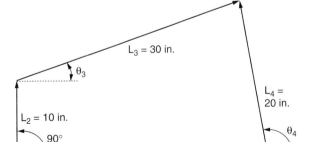

Open a new A-size drawing, and set the scale to 1:5.

Draw three lines at arbitrary orientations, as shown in Figure 10.32, without adding vertical, horizontal, or perpendicular relations.

Select the three lines, and increase their thicknesses from the Line Format toolbar.

FIGURE 10.32

FIGURE 10.33

Add a centerline representing Link 1, and add a Horizontal relation to this line, as shown in Figure 10.33.

Add length dimensions, as shown in Figure 10.34. Be sure to add dimensions oriented with the links, not horizontal or vertical dimensions.

FIGURE 10.34

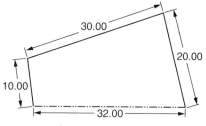

Add the angular dimension between Links 1 and 2, setting it to 90 degrees, as shown in Figure 10.35. Add a Fix relation to one of the points on Link 1.

FIGURE 10.35

Note that although both points on Link 1 are fixed, the 32-inch distance and horizontal relation cause one point to be fully defined when the point at the opposite end is fixed. Therefore, adding a Fix relation to both points would result in the drawing being overdefined.

FIGURE 10.36

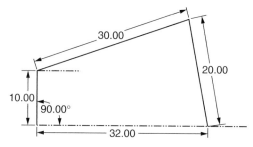

Add horizontal centerlines from which the angular positions of Links 3 and 4 will be measured, as shown in Figure 10.36.

Add an angular dimension between Link 3 and the adjacent horizontal centerline. A box will appear with the message that this dimension will overdefine the drawing. Select OK to make the dimension driven, as shown in Figure 10.37.

FIGURE 10.37

FIGURE 10.38

Add an angular driven dimension defining the position of Link 4. Right-click on each these dimensions and select Display Options: Show Parentheses, as shown in Figure 10.38.

When driven dimensions are shown with parentheses, as shown in Figure 10.39, it makes clear that these dimensions do not control the position of the mechanism. Double-clicking on these driven dimensions will not allow their values to be edited.

While the angular dimensions found from Figure 10.39 could be easily found with any 2-D CAD system, the advantages

FIGURE 10.39

FIGURE 10.40

of the SolidWorks drawing environment is seen when multiple configurations of the mechanism are to be found. For example, if we need to find θ_3 and θ_4 for a value of $\theta_2 = 45$ degrees, we need to change only that dimension.

FIGURE 10.41

Double-click on the 90-degree dimension and change its value to 45 degrees.

Notice that the driven dimensions θ_3 and θ_4 are updated accordingly, as shown in Figure 10.40.

A driving dimension can be changed to a driven dimension by right-clicking and selecting Properties, and then checking the Driven box, as shown in Figure 10.41.

Change the 45-degree dimension to Driven.

We can now click and drag the endpoint of Link 2, as shown in Figure 10.42, to investigate the range of possible motions. This mechanism is classified as a *crank-rocker* because one of the members that pivots about a fixed point can rotate 360 degrees (a crank), while the other member that pivots about a fixed point oscillates back and forth (a rocker). We notice that as the crank, Link 2, revolves, there are two positions for which Link 4, the rocker, is vertical. If we want to define the positions of the crank for which this condition applies, then we need to make θ_4 a driving dimension.

FIGURE 10.42

FIGURE 10.43

Drag one of the links until Link 4 is almost vertical, as shown in Figure 10.43.

Right-click on the dimension defining the angular position of Link 4 and clear the Driven box. Change the dimension to 90 degrees.

We see in Figure 10.44 that θ_2 equal to 66.23 degrees results in Link 4 being vertical.

When dragging Link 2 through its full range of motion, we found that there are two configurations for which Link 4 is vertical.

Set the dimension defining Link 4 to Driven. Drag the mechanism to approximately the position shown in Figure 10.45, and set the dimension back to Driving. Set the dimension to 90 degrees.

FIGURE 10.44

FIGURE 10.45

FIGURE 10.47

We see that $\theta_2 = 357.8$ degrees (or -2.2 degrees) is another solution for which the rocker is vertical.

We can now examine the effect of changing the length of one of the links.

FIGURE 10.46

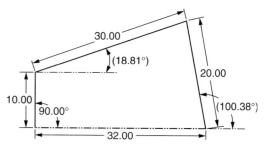

Set the mechanism back to its original position, with θ_2 as the driving dimension and θ_4 as a driven dimension, as shown in Figure 10.46.

Double-click the 30-inch dimension and change its value to 20 inches, as shown in Figure 10.47.

Change the 90-degree dimension to 150 degrees, as shown in Figure 10.48.

FIGURE 10.48

Note that the position of the links remains unchanged, but the links turn red. This is because no solution can be found based on the input dimensions. To find why, we can drag the links to determine their range of motion.

Set the 150-degree dimension to Driven, and drag the endpoint of Link 2 as far as possible in the counterclockwise direction, as shown in Figure 10.49.

FIGURE 10.49

We find that when θ_2 reaches about 138 degrees, Links 3 and 4 are aligned, preventing Link 2 from rotating farther. This condition defines a *toggle position* of the mechanism. The mechanism is called a *double-rocker*, since neither link that pivots around a fixed point can rotate 360 degrees.

To find the precise location of the toggle position, a relation between Links 3 and 4 can be added.

Select Links 3 and 4, and add a collinear relation. The value of θ_2 shown in Figure 10.50, 138.05 degrees, defines the toggle position.

FIGURE 10.50

A second toggle position exists at $\theta_2 = -138.05$ degrees.

If you want to move the mechanism, it is necessary to select either Link 3 or Link 4 and delete the collinear relation from the PropertyManager, or use the Undo key to remove the relation.

Problems

P10.1–10.4   Find the vector **C**, which is the sum of vectors **A** and **B**, graphically. Check your results by adding the x and y components of the vectors. All angles are measured CCW from the +x axis, as shown in Figure P10.1

FIGURE P10.1

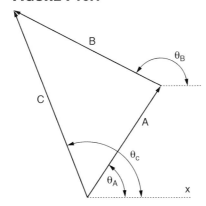

	Vector **A**		Vector **B**	
	Magnitude	Angle	Magnitude	Angle
P10.1	300 lb.	45°	150 lb.	90°
P10.2	10 m/s	0°	15 m/s	270°
P10.3	50 N	−10°	70 N	80°
P10.4	5 ft./s²	120°	4 ft./s²	300°

P10.5–10.8   Consider the vector equation **A** + **B** = **C**. Graphically find the unknown quantities for each set of vectors. All angles are measured CCW from the +x axis.

	Vector A		Vector B		Vector C	
	Magnitude A	Angle θ_A	Magnitude B	Angle θ_B	Magnitude C	Angle θ_C
P10.5	200 lb.	25°	?	160°	300 lb.	?
P10.6	?	?	22 m/s	90°	36 m/s	78°
P10.7	?	130°	?	95°	50 N	110°
P10.8	8 ft./s²	112°	5 ft./s²	?	?	88°

P10.9 The mechanism illustrated here is called an *offset slider-crank.*

As Link 2, the crank, rotates, Link 4, the slider, moves back and forth along a horizontal line. The distance L_3 is the horizontal distance from the pivot point of the crank to the pin joint between the connector, Link 3, and the slider. Link 1 is the ground, and the distance L_1 is the offset distance.

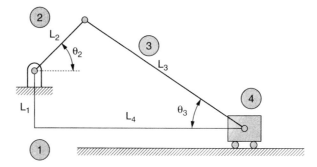

Construct a layout drawing of the mechanism, with $L_1 = 30$ mm, $L_2 = 50$ mm, and $L_3 = 150$ mm.

a. Find L_4 and θ_3 for a crank angle $\theta_2 = 45$ degrees
    (Answers: 170.4 mm, 25.8 degrees)
b. Find L_4 for values of θ_2 of 0, 90, 180, and 270 degrees.
c. Find the minimum and maximum possible values of L_4, and the corresponding crank angles.

Using SolidWorks in the Design and Analysis of Mechanisms

Introduction

A *mechanism* is an assemblage of *links* and *joints* that are connected together to achieve a desired motion task. The *links* provide the mechanical structure of the mechanism, while the *joints* provide the ability for the mechanism to move. Some typical kinds of mechanisms used by mechanical engineers are shown in Figure 11.1.

The ways in which we can connect links with joints and provide a mechanism with the ability to move are seemingly endless! Also, the links can take on any shape and size we desire, and the motion may change in complicated and seemingly unpredictable ways as we modify the links. The design of a mechanism may seem like a daunting task.

Computer-aided design (CAD) packages have become valuable tools in the design and analysis of mechanisms. The ability for the design engineer to adjust the size, shape, and interconnectivity of links and joints and quickly assess the impact on the mechanism has accelerated the design cycle of mechanisms. The SolidWorks program, with its ability to represent geometric constraints between structural components using assembly mates, is ideally suited to the design and virtual prototyping of complex mechanisms.

The remainder of this chapter will be devoted to the use of the SolidWorks program in the design of mechanisms through a case study involving the design of a *four-bar linkage*.

FIGURE 11.1

Four-bar linkage

Six-bar linkage

Slider-crank

In this chapter, you will:

- learn some basic terminology associated with mechanism design,

- use the assembly modeling techniques learned in Chapter 6 to develop a working model of a mechanism, and

- perform a simple motion analysis of a mechanism.

219

11.1 Approaching Mechanism Design with SolidWorks Assemblies

Consider the *four-bar linkage*, a classic mechanism used in engineering, shown in Figure 11.2.

It consists of three structural *links*, connected to each other and to fixed pivot points by *pin joints* that allow for rotating motion between the links. Though it only has three physical links, it is called a *four-bar linkage* because there is an implied fourth structural link (or *bar*) that connects the fixed ground points, as shown in Figure 11.3.

FIGURE 11.2

FIGURE 11.3

While the choice of a four-bar linkage as our preliminary design solution is an important step, the *parametric design problem* of selecting the appropriate link lengths and ground pivot locations to give us the desired motion is a difficult engineering task. We will develop a SolidWorks model to aid in this parametric design phase.

The features that we will employ are the *assembly* capabilities. The development of assembly models and the definition of assembly mates were covered in Chapter 6. Mated assemblies are an ideal tool for use in mechanism design, since the *joints* that provide the physical relationships between the *links* are analogous to the *mates* that define the geometric relationships between *parts*.

Think about two links connected by a pin joint, as shown in   **FIGURE 11.4**
Figure 11.4.

The pin through the holes in the links allows for rotation between the links, giving a "scissors" action. The *mated assembly* representation of this type of link/joint assembly would involve two parts, with two mates serving the same geometric purpose as the pin joint in the physical linkage:

- The front face of one link is aligned with the back face of the other link using a *coincident mate*; this allows one link to "slide by" the other without any interferences between the surfaces.

- The hole in one link is aligned with the hole in the other link using a *concentric mate*; this keeps the holes aligned at their central axes.

By capturing the essential geometric constraints that underlie a *pin joint*, this sequence of two mates imparts the same motion capabilities to the solid model that are seen in the physical mechanism.

Using this simple two-link mechanism as a building block, complex mechanisms can be assembled and virtually tested. Mechanism motion can be tested and debugged without the need for physical prototypes to be constructed. The following section will step through the development of a model of a four-bar linkage.

11.2 Development of Part Models of Links

In this section, a step-by-step tutorial will lead you through the development of the part models required to make a "working" assembly model of a four-bar linkage. Four links, similar in shape but of different dimensions, will be constructed.

FIGURE 11.5

Open a new part, and select the Front Plane. Drag a rectangle from the origin, as shown in Figure 11.5.

Select the Tangent Arc Tool. Click on a corner point of the rectangle, and drag the arc away from the rectangle, as shown in Figure 11.6.

FIGURE 11.6

FIGURE 11.7

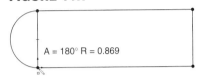

Drag the arc so that the endpoint snaps to the corresponding corner of the rectangle, as shown in Figure 11.7.

Draw a second tangent arc connecting the endpoints (Figure 11.8).

FIGURE 11.8

FIGURE 11.9

Use the Trim Entities Tool to remove the short sides of the rectangle, as shown in Figure 11.9.

Now add the dimensions shown in Figure 11.10.

FIGURE 11.10

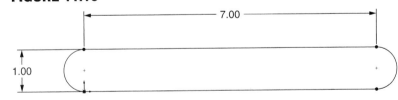

`   The sketch should be fully defined.

FIGURE 11.11

Extrude the sketch to a depth of 0.25 inches (Figure 11.11).

Select the front face of the part and choose the Circle Tool.

Two circles, concentric with the arcs at the ends of the link, will now be added. In order to activate the center points of the arcs as snap points, these features must first be "woken up."

FIGURE 11.12

"Wake up" one of the center marks by momentarily holding the cursor on the circular edge of the part (Figure 11.12).

The center point of the arc will now be available as a snap point.

FIGURE 11.13

R = 0.29

Sketch a circle centered at this snap point, as shown in Figure 11.13. Repeat this process at the other end of the link. Press the esc key to turn off the Circle Tool. Select both circles, and add an equal relation. Dimension one of the hole diameters as 0.50 inches, as shown in Figure 11.14.

The sketch should be fully defined.

FIGURE 11.14

Ø.50

Extrude a cut (type Through All) to create the holes.

The finished part is shown in Figure 11.15.

Save the part as "Link1."

The easiest way to create each of the other three links is to modify the length of the first part and save it with a different name. All dimensions except the length should be identical for the four links.

FIGURE 11.15

FIGURE 11.16

1.00

7.00

.25

Double-click Extrude1 in the FeatureManager. The dimensions defining this feature will be displayed, as shown in Figure 11.16.

Double-click the 7-inch length and change it to 3 inches. Rebuild the model (Figure 11.17).

Choose File: Save As from the main menu. Save this part as "Link2."

FIGURE 11.17

Create the other two links, using the lengths as shown in Table 11.1, and save them using the names "Link3" and "Link4." Close the part window.

Table 11.1 Lengths of the Four Link Parts

	Length
Link1	7 in.
Link2	3 in.
Link3	5 in.
Link4	6 in.

11.3 Development of the Assembly Model of the Four-Bar Linkage

The part models developed in Section 11.2 will now be used to construct an assembly model of the four-bar linkage. It is this assembly model that will allow us to perform parametric design and simple motion analysis of the mechanism.

Open a new assembly. If the dialog box shown in Figure 11.18 does not appear, open it by selecting the Insert Component Tool from the Assembly group of the CommandManager. Select Browse, and select the file Link1.

Click anywhere in the drawing space to place this part.

FIGURE 11.18

FIGURE 11.19

This first part will be fixed in space—note the "(f)" beside its name in the Design Tree. We refer to it as the "ground link" since all motion of the other links will be relative to this fixed link.

Select the Insert Component Tool. Browse to select the file Link2. Click to place Link2 in the approximate position shown in Figure 11.19.

FIGURE 11.20

Note that you can click and drag Link2 to move it to a new position, as shown in Figure 11.20.

A coincident mate aligning the faces of these links will be defined.

Select the Mate Tool to initiate a new mate.

The Mate PropertyManager will open.

Using the Rotate View Tool, rotate the parts so that the back faces of the links can be seen. Select the back face of Link2, as shown in Figure 11.21.

FIGURE 11.21

Switch to the Trimetric View, and select the front face of Link1, as shown in Figure 11.22.

Link2 will automatically move toward Link1 so that the selected faces show a coincident mate.

FIGURE 11.22

Click the check mark in the PropertyManager or the pop-up box to apply the mate.

FIGURE 11.23

A concentric mate between the holes must also be added to simulate the kinematic constraint of the pin joint.

Select the inside surfaces of the holes at the left end of the links (Figure 11.23).

FIGURE 11.24

A concentric mate will be previewed, as shown in Figure 11.24. Click the check mark to apply the mate. Click the check mark again to close the Mate PropertyManager.

The kinematic constraints of the pin joint are now fully defined. Link2 is still free to move, as long as the movements do not violate the mates that we have placed on it. We can experiment to see the type of motion still allowed under the constraints we have imposed.

Click and drag on Link2. Confirm that you can rotate the link though a full 360 degrees. Leave Link2 in the approximate position shown in Figure 11.25.

FIGURE 11.25

Note that the only unconstrained motion is rotation about the mated hole, as if the links were pinned together.

Select the Insert Component Tool. Select Link3 and place it into the assembly.

Select the Mate Tool.

In earlier chapters, we learned a shortcut method of selecting faces not visible from the current view orientation. We will use that technique to select the back face of Link3.

Move the cursor over Link3. Right-click, and pick Select Other, as shown in Figure 11.26.

FIGURE 11.26

The back face of the link is highlighted, as shown in Figure 11.27. Click the left mouse button (Y) to accept the selection. Select the front face of Link1. A coincident mate will be previewed; click the check mark to accept the mate.

Add a concentric mate to the inner faces of corresponding holes in Link1 and Link2. Close the Mate PropertyManager.

FIGURE 11.27

Click and drag Link3 to the approximate position shown in Figure 11.28.

Insert Link4 into the assembly. Add three mates:

1. **a coincident mate between the back face of Link4 and the front face of either Link2 or Link3,**

2. **a concentric mate between the inner faces of the corresponding holes of Link2 and Link4, and**

3. **a concentric mate between the inner faces of the corresponding holes of Link3 and Link4.**

FIGURE 11.28

Note that only one coincident mate is required. The coincident mate between the back face of Link4 and the front face of one of the other links completely defines the z-direction location of Link4. If another coincident mate is added between the back face of Link4 and the front face of another link, an error will be reported, as this mate will overdefine the assembly.

The completed assembly is shown in Figure 11.29.

While we can rotate Link2 manually, we can also add a simulated motor and animate the motion of the assembly.

Select the Simulation Tool from the Assembly group of the CommandManager. Pick Rotary Motor from the pop-up menu, as shown in Figure 11.30.

Click on the front face of Link2, as shown in Figure 11.31. This places a motor driving Link2 around an axis perpendicular to the selected face.

In the PropertyManager, the relative velocity of the motor can be set with a slider bar, as shown in Figure 11.32. Leave the velocity setting near the middle, and click the check mark to close the PropertyManager.

Select the Simulation Tool, and select Calculate Simulation from the pop-up menu, as shown in Figure 11.33.

FIGURE 11.29

FIGURE 11.30

FIGURE 11.31

FIGURE 11.32

FIGURE 11.33

FIGURE 11.34

The linkage will continuously rotate through its range of motion, as illustrated in Figure 11.34.

Machine Dynamics and Machine Design

In this chapter, we have examined a four-bar linkage and performed some simple motion analysis. If we were designing this mechanism for an engineering application, many more questions would remain:

- How fast will the output link oscillate, if we know the speed of the input link?

- What size motor would be required to drive the input link?

- How will we transmit the rotational power of the motor to the input link?

Providing the answers to these questions would generally require the expertise of a mechanical engineer. The analysis of link speeds and accelerations is classified as a kinematics problem; extending this analysis to the sizing of motors is classified as a kinetics problem. Most engineering programs include basic courses in physics and dynamics, which address the essentials of kinematic and kinetic analysis. However, because of the great emphasis in mechanical engineering on applying these principles to mechanisms and machinery, many mechanical engineering programs include upper-level courses in the advanced application of kinematic and kinetic analysis. Such a course is often called a *machine dynamics* course, and may include the study not only of linkages, but also of the dynamics of gears, cams, and other mechanical devices.

The question of power transmission requires further insight beyond kinematic and kinetic analysis. The choice of transmission also requires the investigation of various alternatives, such as a geared transmission, a system of belts and pulleys, and a chain and sprocket drive. The selection and sizing of the appropriate transmission system requires knowledge not only of kinematic and kinetic analysis, but also of the application of stress analysis to the transmission components. This type of analysis is often covered in mechanical engineering programs in *machine components* or *machine design* courses.

To end the simulation, select the Simulation Tool and Stop Record or Playback, as shown in Figure 11.35.

To place the links into the position in which they were assembled, select the Simulation Tool and choose Reset Components, as shown in Figure 11.36.

Save the assembly file.

The behavior of the assembly, with Link2 rotating through a full 360 degrees, is controlled by the relative lengths of the links. In particular, there will always be one link that can rotate fully (this link is called a "crank") if this equation is satisfied:

$$L + S < P + Q$$

where:

L = the length of the longest link

S = the length of the shortest length

P and Q = the lengths of the other two links.

FIGURE 11.35

FIGURE 11.36

For our linkage, L = 7 inches, S = 3 inches, and P and Q are 5 and 6 inches. Therefore, the condition above is satisfied, and at least one link (Link2) is a crank.

The SolidWorks program allows you to easily modify one of the links, and the associativity between parts and assemblies causes the changes you make to your parts to be reflected in the assembly. In fact, parts can be edited directly from the assembly window.

Double-click on Link4. The dimensions associated with the part are displayed, as shown in Figure 11.37. Double-click the 6-inch dimension and change it to 4 inches. Rebuild the assembly.

FIGURE 11.37

The modified assembly is shown in Figure 11.38. If we consider the equation discussed above, we find that L is still 7 inches, while S remains 3 inches. P and Q now equal 4 and 5 inches. Therefore, L + S is greater than P + Q, so the condition for a crank does not exist.

FIGURE 11.38

Select the Simulation Tool, and choose Calculate Simulation.

The linkage begins to move as before, but "locks up" when it reaches a certain point, as shown in Figure 11.39.

FIGURE 11.39

FIGURE 11.41

Select Simulation: Stop Record or Playback.

From the front view, we see that the motion has stopped when Link3 and Link4 are in alignment, as shown in Figure 11.40.

FIGURE 11.40

In the FeatureManager, right-click on the RotaryMotor (you may need to click the + sign next to the Simulation item to expand it), and choose Edit Feature. Click the Reverse Direction icon in the PropertyManager, as shown in Figure 11.41.

Run the Simulation.

This time the linkage will lock up in the position shown in Figure 11.42.

Stop the simulation.

FIGURE 11.42

The positions in which the mechanism "locks up" are called *toggle positions*. (Recall that in Chapter 10, we used 2-D layouts to determine the precise orientations of the toggle positions.) The presence of toggle positions may be undesirable, since they prevent full rotation of the input link. In other cases, they may be desirable, if the mechanism is to be used as a clamp or fixturing device.

While we have looked at mechanisms for which L + S is less than P + Q (one link can rotate 360 degrees), and for which L + S is greater than P + Q (no link can rotate 360 degrees), it is logical to ask what will happen if L + S is *exactly equal to* P + Q.

FIGURE 11.43

Click and drag Link2 into the approximate position shown in Figure 11.43.

Double-click Link4 to display its dimensions. Double-click the 4-inch length and change it to 5 inches. Rebuild the assembly.

The values of L and S remained unchanged at 7 and 3 inches, respectively. The values of P and Q are 5 inches each, so L + S = P + Q.

Run the simulation.

The motion of the linkage will appear unusual. If you watch it carefully, you will notice that for two complete *cycles* (360-degree rotations) of Link2, the motions of Link3 and Link4 are different from one cycle to the next. In Figure 11.44, positions from one cycle are shown, while in Figure 11.45, positions for the next cycle are shown.

FIGURE 11.44

FIGURE 11.45

This mechanism is *unstable*. When the linkage passes through a certain point (when the links are all lying flat), there are two possible paths for the links to take. In a stable four-bar mechanism, the links can be assembled into either the open configuration (Figure 11.46) or the crossed configuration (Figure 11.47). Physically, the only way to

FIGURE 11.46   **FIGURE 11.47**

change from one configuration to another is to remove the pin connecting two of the links, move the links to the other configuration, and replace the pin. The unstable mechanism switches from the open to the crossed configuration. When used in a real application to drive a moving component, the behavior of such a mechanism may be unpredictable.

Close the assembly window without saving the changes.

Problems

P11.1   A type of mechanism used in engineering systems is the *slider-crank* (see Figure P11.1a). In the slider-crank, the input link (crank) rotates continuously through a full 360 degrees, while the output slider slides along a fixed surface. Among other things, the *slider-crank* is the working schematic for a single cylinder of an internal combustion engine. Create a working SolidWorks assembly model of this mechanism, using the dimensions shown in Figure P11.1b. The dimensions are mm. (Hint: The bottom of the slider is always coincident with the ground plane.)

FIGURE P11.1a   **FIGURE P11.1b**

P11.2   The mechanism described in P11.1 can be modified by offsetting the pivot location of the crank from the path of the slider, creating an *offset slider-crank*, as shown in Figure P11.2a. This type of mechanism is sometimes called a *quick-return* mechanism, since the back-and-forth motion of the slider is different in one direction from the other. Create a working model of this mechanism, using the dimensions shown in Figure P11.2b. The dimensions are mm.

FIGURE P11.2a   **FIGURE P11.2b**

P11.3   Additional links can be added to a four-bar linkage to modify the motion of the links. A common engineering example is the *six-bar linkage*. Figure P11.3a shows a model of a six-bar linkage that simulates the operation of a car's windshield wipers. The Crank is attached to the wiper motor, which rotates at a constant speed. The lengths of the Crank and Connector 1, along with the attachment location to the blade, are selected so that the rotational motion of the wiper motor is converted to the desired oscillating motion of the first wiper blade. Connector 2 links the two blades together so that they move parallel to each other.

FIGURE P11.3a   **FIGURE P11.3b**

FIGURE P11.3c

Model the components shown in Figure P11.3b, and assemble them into the mechanism, using the spacing of ground joints shown in Figure P11.3c.

P11.4   In the 1700s, noted engineer James Watt devised and patented a mechanism for generating straight-line motion with a rotational input. In this mechanism, shown in Figure P11.4a, point X will trace out a straight line in space as the input link (Link 2) rotates, as illustrated in Figure P11.4b. Originally designed for guiding the stroke in a steam engine piston, this mechanism is currently used to guide axle motion in automotive suspension applications[1]. Develop a working model of the *Watt Straight Line Mechanism*. Use a length of 8 inches for Links 2 and 4, and a length of 4 inches for Link 3. Point X is in the center of Link 3. The distance between the fixed pivot points is 16 inches. (Note: This a considered a *double rocker mechanism*; the input link is *not* able to rotate through a full 360 degrees.)

FIGURE P11.4a

FIGURE P11.4b

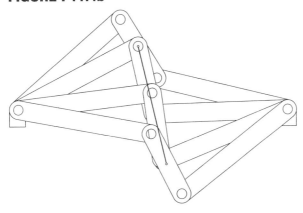

P11.5   A *geneva mechanism*, which is illustrated in Figure P11.5a, is used to transform constant rotational motion into intermittent motion. Among other uses, this type of mechanism is used to control the motion of an *indexing table* in an assembly line. An indexing table will remain stationary for a period of time, and then periodically rotate a fixed amount.

FIGURE P11.5a

[1]Robert L. Norton, *Design of Machinery*, 2nd ed. (New York: McGraw-Hill, 1999).

Model the components necessary to create a model of the geneva mechanism:

a.   The *geneva wheel*, as detailed in Figure P11.5b (thickness = 0.25 inches)

FIGURE P11.5b

b.   The *locking disk*, as detailed in Figure P11.5c (thickness = 0.25 inches)

FIGURE P11.5c

c.   The *crank*, as detailed in Figure P11.5d (thickness = 0.25 inches)

FIGURE P11.5d

d.   The *ground link*, as shown in Figure P11.5e (thickness = 0.25 inches)

FIGURE P11.5e

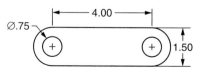

e.   A cylindrical *shaft*, 0.75 inches in diameter by 2 inches long

f.   A cylindrical pin, 0.375 inches in diameter by 0.50 inches long.

Create the subassembly shown in Figure P11.5f from a shaft and the geneva wheel. Add mates so that the wheel is completely fixed relative to the shaft.

FIGURE P11.5f

Create the subassembly shown in Figure P11.5g from a shaft, the crank, the pin, and the locking disk. Add mates so that the crank, pin, and disk are all completely fixed relative to the shaft.

FIGURE P11.5g

Create a new assembly with the ground link as the fixed component. Add the two subassemblies created above, and create mates so the shafts can rotate within the ground link. Add a rotary motor to the crank/locking disk assembly, and watch the motion, which is illustrated in Figure P11.5h.

FIGURE P11.5h

Notes: Make sure that the initial orientation of the locking disk is such that one of the cutouts of the wheel is aligned so that the locking disk can be placed without interference. When the motion is simulated, the simulation may slow down slightly when the pin enters the slot of the wheel and the contact of the parts is recognized. If you replay the simulation, the motion of the assembly will be smooth.

The Use of SolidWorks as a Tool for Manufacturing: Mold Design and Sheet Metal Parts

Introduction

A design engineer must always consider the method of manufacture when designing any part. Failure to do so may result in part designs that are more expensive to make than necessary, have high scrap rates, or cannot be made at all.

Some manufacturing processes create unique challenges from a solid modeling standpoint. For example, when designing a mold, the shape of the part to be molded must be removed from the interior of the mold, usually with the dimensions adjusted to allow for shrinkage of the part during the cool-down portion of the molding cycle. Sheet metal parts are cut from flat material, and then bent into the final shape. Therefore, the part definition must include both the flat shape and the finished geometry. The SolidWorks program has specialized tools for working with molds and sheet metal parts.

In this chapter, you will:

- create a cavity within a mold base,
- create and modify two mold halves that are linked to the mold base with a cavity, and
- make a simple sheet metal part and learn how to show the part in either the flat or bent state.

12.1 A Simple Two-Part Mold

In this exercise, we will make a mold to produce a simple part, a cylinder. We will begin by making the part itself.

Open a new part. Draw and dimension a 3-inch diameter circle in the Right Plane, centered at the origin. Extrude the circle using a midplane extrusion, as shown in Figure 12.1. Set the total depth of extrusion to 5 inches. Save this part with the file name "Cylinder." Do not close the part window.

FIGURE 12.1

The next step is to create a *mold base,* from which a *cavity* the shape of the part will be removed. The mold base must be large enough to completely enclose the part.

Open a new part. Draw and dimension a 4-inch-by-6-inch rectangle in the Top Plane. Add a centerline between two opposite corners of the rectangle. Add a Midpoint relation between the centerline and the origin, as shown in Figure 12.2.

FIGURE 12.2

Extrude the rectangle with a midplane extrusion, as shown in Figure 12.3. Set the total depth of extrusion to be 4 inches. Save this part with the file name "Base." Do not close the part window.

FIGURE 12.3

FIGURE 12.4

We will now place the part within the mold base.

Select the Make Assembly Tool from the Standard toolbar, as shown in Figure 12.4. Click OK to accept the default assembly template.

A new assembly window will open, and the base will appear as the first component to be inserted.

Click on the origin of the assembly to place the base, as shown in Figure 12.5.

FIGURE 12.5

Note that placing the origin of the first component of an assembly at the origin of the assembly itself is not required, since the first component will be fixed and subsequent parts will be located relative to the first part. However, placing the first part in this manner is good practice, since the Front, Top, and Right Planes of the assembly can be used as mating entities.

Select the Insert Component Tool. Select the cylinder part from the list of open files, as shown in Figure 12.6, and click to place it in the assembly, as shown in Figure 12.7.

Select the Mate Tool.

FIGURE 12.6

FIGURE 12.7

FIGURE 12.8

In the FeatureManager, which flies out to the right of the Property-Manager, click on the + sign to each of the parts to expand their definitions, if necessary. Select the Front Plane of each, as shown in Figure 12.8.

A preview of the resulting coincident mate is shown in Figure 12.9.

Click the check mark to complete the mate.

Repeat this procedure to mate the Top Planes, and then the Right Planes, of the base and cylinder.

Save this assembly with a file name of "Mold Assembly."

FIGURE 12.9

FIGURE 12.10

The cylinder is now centered within the base, as shown in the wireframe view of Figure 12.10. Rather than work in a wireframe mode, it is helpful to display the base as transparent.

Click on the part name "Base" in the FeatureManager to select it. Click the Edit Color Tool. Near the bottom of the PropertyManager, select the transparency slider bar, as shown in Figure 12.11. Move this bar to the right to display the base in varying degrees of transparency. Click the check mark to apply the desired transparency.

FIGURE 12.11

FIGURE 12.12

The cylinder can now be seen within the base, as shown in Figure 12.12.

We will now create the cavity. Since the cavity will be created in the base, the base must be selected for editing.

Select the base from the FeatureManager. Choose the Edit Component Tool from the Assembly group of the CommandManager.

Edit Compo...

From the main menu, select Insert: Molds: Cavity, as shown in Figure 12.13.

FIGURE 12.13

In the PropertyManager, select the Cylinder part from the FeatureManager as the component defining the cavity. Set the "Scale about" option to "Component Centroids" and the scale to 3%, as shown in Figure 12.14. Click the check mark to create the cavity. Click on the Edit Component Tool to end the editing of the base.

The scale factor causes the cavity to be larger than the finished part. Most molding materials shrink during cure or cooling, so the scale factor allows for that shrinkage.

The cavity can now be seen within the base, as shown in Figure 12.15. If you look closely, you can see that there are gaps between the edges of the part and the corresponding edges of the cavity, because of the shrink factor.

The base, which now includes the cavity, needs to be split into two halves. For this simple mold, the two halves will be identical, and so we could cut away half of the original base part. However, if the two mold halves will be different (as in the next exercise), then copies of the base must be made. This procedure of creating *derived parts* is illustrated here.

FIGURE 12.14

FIGURE 12.15

FIGURE 12.16

FIGURE 12.17

Select the Base from the FeatureManager. From the main menu, select File: Derive Component Part, as shown in Figure 12.16.

A new part window is opened, and a copy of the base (including the cavity), is created, as shown in Figure 12.17. The advantage of creating a derived part rather than simply saving a copy of the base part is that associativity is maintained. That is, if a change is made to the cylinder part, then the cavity in the assembly, the base, and the derived mold half part are all updated.

FIGURE 12.18

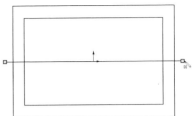

Select the front face of the new part. Select the Line Tool, and draw a line completely through the part, passing through the origin, as shown in Figure 12.18. Extrude a cut with a type of Through All, with the direction to cut arrow pointing up.

FIGURE 12.19

The resulting part is shown in Figure 12.19.

Save the part with a file name of "Mold Half."

FIGURE 12.20

To illustrate the associativity of the parts, open the cylinder file and change the diameter of the part from 3 to 2 inches. If you switch immediately to the mold half part, then no changes will be seen. However, if you switch to the mold assembly, then the cavity will be updated. Then, switching to the mold half will cause the change to be reflected, as shown in Figure 12.20.

Two of these mold halves can now be assembled, and mold-level features (fill and vent ports, alignment pins, etc.) can be added.

12.2 A Core-and-Cavity Mold

In this exercise, a two-piece mold for making the card holder from Chapter 5, shown in Figure 12.21, will be created. The shape of this part requires that the mold geometry consist of a *core* half (Figure 12.22), with features protruding outward from the parting line, and a *cavity* half (Figure 12.23), with features cut inward from the parting line.

FIGURE 12.21

FIGURE 12.22

FIGURE 12.23

Our procedure will be similar to the one followed in the previous section. Since there are several files involved in this procedure, it is helpful to consider the process steps, which are illustrated in Figure 12.24. An interim assembly will be made from the mold base and part, so that the cavity can be placed in the mold base. From that assembly, copies of the mold base (with the cavity) will be derived. These copies will be modified to become the two mold halves. Finally, the two mold halves will be brought together into an assembly.

FIGURE 12.24

The first step of the process will be to create the mold base, which must be sized so that the card holder part fits completely within its boundaries.

Open a new part. In the Top Plane, draw and dimension a rectangle, as shown in Figure 12.25. Use a centerline and a Midpoint relation to center the rectangle about the origin.

FIGURE 12.25

FIGURE 12.25

The dimensions shown are each 1 inch greater than the corresponding dimensions of the card holder, allowing for $\frac{1}{2}$ inch clearance between the part and the mold edges.

The height of the mold base will be selected to allow for $\frac{1}{4}$ inch above and below the part.

Extrude the rectangle upward a distance of 1.5 inches, as shown in Figure 12.26.

FIGURE 12.26

FIGURE 12.27

Choose the Edit Color Tool, and move the Transparency slider bar toward the right, as shown in Figure 12.27.

The transparent part is shown in Figure 12.28.

FIGURE 12.28

Save this part as "Mold Base."

Choose the Make Assembly Tool from the Standard toolbar. Insert the mold base into the Assembly. Choose the Insert Component Tool and browse to locate the card holder part. Click to place it in the assembly in the approximate position shown in Figure 12.29.

FIGURE 12.29

We will now align the card holder with the mold base.

FIGURE 12.30

Rotate the view so that the bottom surfaces of the card holder and the mold base are visible, as shown in Figure 12.30.

FIGURE 12.31

Choose the Mate Tool. Select the bottom surface of the card holder (Figure 12.31), and then the bottom surface of the

Mate

mold base. A coincident relation will be previewed. Select the Distance Mate icon from the PropertyManager (Figure 12.32). **Set the offset distance between the two surfaces to be 0.25 inches. If necessary, check the Flip**

Dimension box so that the bottom surface of the card holder is above the bottom surface of the mold base, as previewed in Figure 12.33. Click the check mark to apply the mate.

FIGURE 12.33

FIGURE 12.32

FIGURE 12.34

Select the Front Planes of the card holder and mold base, as shown in Figure 12.34. Add a coincident mate, as shown in Figure 12.35.

FIGURE 12.35

Add a coincident mate between the Right Planes of the card holder and mold base. Close the Mate PropertyManager.

FIGURE 12.36

The card holder is now placed within the mold base, as shown in Figure 12.36.

Save the assembly as "InterimMoldAssy."

 Select the mold base from the FeatureManager to select it. Choose the Edit Component Tool from the Assembly group of the CommandManager.

FIGURE 12.37

From the main menu, select Insert: Molds: Cavity. In the PropertyManager, select the card holder as the component defining the cavity. Set the "Scale about" option to Component Centroids and the scale to 1%, as shown in Figure 12.37. Click the check mark to create the cavity. Click the Edit Component Tool to end the editing of the mold base.

Select the mold base from the FeatureManager. From the main menu, select File: Derive Component Part.

A new part window is opened, and a copy of the mold base is created, as shown in wireframe mode in Figure 12.38.

We will now cut away the top portion of the mold to create the lower mold half. This cannot be done with a simple extruded cut, since that would also cut through the core feature that will form the underside of the card holder. Rather, we will perform two separate cutting operations to achieve the desired geometry.

FIGURE 12.38

Select the top surface of the part and choose the Sketch Tool from the Sketch group of the CommandManager to open a sketch. Select the four edges of the cavity shown in Figure 12.39. (Remember to use the ctrl key to select multiple entities.)

Sketch

FIGURE 12.39

Select the Convert Entities Tool, which will create lines from the selected edges.

Convert Entities

Extrude a cut with a type of "Up to Next." This will cut away the top of the mold, but only to the cavity, as shown in Figure 12.40.

In order to cut away the material around the edges of the mold half, it is necessary to use a different type of cut: one that cuts down to the parting surface.

FIGURE 12.40

Open a new sketch on the top surface. Convert the eight edges shown in Figure 12.41 into lines, using the Convert Entities Tool.

Extrude a cut, with a type of "Up to Surface." For the surface, choose the bottom surface of the cavity, which corresponds to the parting line, as shown in Figure 12.42.

FIGURE 12.42

FIGURE 12.41

The completed mold half is shown in Figure 12.43.

Save this part as "MoldHalfLower."

We will create the upper mold half.

Switch back to the assembly containing the mold base and the card holder (InterimMoldAssy). Select the mold base, and from the main menu, select File: Derive Component Part.

In the new part, select the front surface, as shown in Figure 12.44. Select the Line Tool from the Sketch group of the CommandManager.

Without any mouse clicks, hold the cursor over the bottom edge of the cavity momentarily, as shown in Figure 12.45.

This will "wake up" the feature and allow the endpoint of the line to be aligned with this edge.

Move the cursor past the left of the mold base, aligned with the bottom edge of the cavity, as shown in Figure 12.46.

Click and drag a horizontal line across the mold base, as shown in Figure 12.47. Select the Extruded Cut Tool from the Features group of the CommandManager.

Since we are not using a closed sketch, we must define whether the material above or below the line is to be cut away.

Check the "Flip side to cut" box in the PropertyManager, so that the arrow indicating the part to be cut away is pointing down, as shown in Figure 12.48.

Before completing the cut, rotate the view orientation so that you can see the bottom of the mold, as shown in Figure 12.49. Click the check mark.

FIGURE 12.43

FIGURE 12.44

FIGURE 12.45

FIGURE 12.46

FIGURE 12.48

FIGURE 12.47

FIGURE 12.49

Part # 15122

Since this cut will produce two separate solid bodies (the mold half and part of the core), you are prompted to identify which of the bodies you want to keep. In our case, we do not want to keep the core with the upper mold half. In the dialog box, checking the Selected bodies box allows you to choose which of the solid bodies you want to keep, and which will be deleted.

FIGURE 12.50

Check the Selected bodies box. Check the box for Body 1 (the mold half) and clear the box for Body 2 (the core). The selected bodies are highlighted in green, as shown in Figure 12.50.

FIGURE 12.51

The resulting part is shown in Figure 12.51.

Save this part as "MoldHalfUpper."

FIGURE 12.52

Open a new assembly. Insert the two mold halves, as shown in Figure 12.52.

Add three mates to align the mold halves together, as shown in Figure 12.53.

FIGURE 12.53

Choose the Section View Tool from the View toolbar.

Click the check mark to accept the Front Plane as the section plane, as shown in Figure 12.54.

FIGURE 12.54

The mold cavity can now be clearly seen.

Save the file with the name "Mold Assembly."

Suppose that we now want to change the thickness of the molded part. Often, if a material change is specified, the thickness will need to be changed, since the flow of the material in the mold is a limiting factor on the thickness.

Open the card holder part. Right-click on the shell feature in the FeatureManager. Select Edit Feature, and double the thickness, from 0.06 to 0.12 inches. Rebuild the model.

The modified part is shown in Figure 12.55.

FIGURE 12.55

Open, or switch to, the interim mold assembly, then the two mold halves, and finally the mold assembly. At every step, the models will be updated to reflect the new thickness.

The updated mold assembly is shown in Figure 12.56. Of course, the mold we created is not usable without some way to get material into the mold. The manner in which this is done depends on the molding process, as discussed in the Future Study box. Many other features may also be required for the mold to be usable. Mold design is a very specialized field, combining mechanical design with material science. However, an essential part of any mold design is the creation of the mold cavity and the separation of the core and cavity mold parts, such as we have done in these exercises.

FIGURE 12.56

FUTURE STUDY

Materials and Processes

Our simple two-piece mold includes a cavity the shape of the finished part. If we created this mold for making a few prototype parts from a room-temperature curing material, such as polyurethane, then we could add holes for bolting the halves together, a hole for pouring in the material, and a vent hole, and our mold would be functional.

Most molds require many other features, however, and tooling design is an important function at any manufacturing company. If a plastic part is to be injection-molded, then the injection points and vent locations must be carefully designed so that the molten plastic fills the cavity completely. The plastic's melt and cooling temperatures and its resistance to flow must be considered when designing both the part and the mold. The tolerances required for the finished part might require that a filler be added to the material for dimensional stability. Ejector pins might need to be added to help remove the part from the mold. (Note that in our example, even though the part walls are tapered, the shrinkage of the

part onto the core will result in forces that will need to be overcome in order to remove the part from the lower mold half.) Since the plastic must be injected hot but allowed to cool before removal from the mold, cooling lines for circulating water are usually added to the mold halves.

Other materials have different processing requirements. Composite materials used for automotive materials are mostly compression molded, in which the raw material is placed between two mold halves and formed by applying pressure with a hydraulic press. Lower-quantity parts can be produced by resin transfer molding, in which dry fabric is placed in a mold and liquid resin is pumped in under low pressure.

Often the choice of a process depends on the quantity of parts to be made. A high-quality tool for injection molding can cost tens of thousands of dollars, but if this cost can be spread over 100,000 or 1,000,000 parts, then the fast cycle times resulting from a good mold design can result in significant cost savings.

12.3 A Sheet Metal Part

FIGURE 12.57

In this exercise we will create the sheet metal part shown in Figure 12.57. The part can be shown in either the bent or flat state.

Open a new part. From the main menu, select Tools: Options: Document Properties: Units. Set the unit system to MMGS (millimeters), and the number of decimal places to zero for both the length and angular values.

Select the Right Plane. Sketch and dimension the three lines shown in Figure 12.58. Use the vertical centerline to align the endpoints of the diagonal lines.

FIGURE 12.58

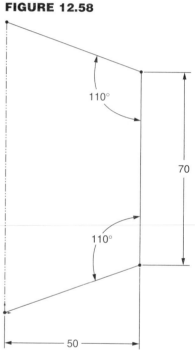

Note that we are not adding radii to the sharp corners. In sheet metal parts, *bends* are added as separate features to a part.

Select the Extruded Boss/Base Tool from the Features group of the CommandManager. Set the extrusion depth to 100 mm. Since the sketch is open, a thin-feature extrusion will be created. Set the thickness to 2 mm, as shown in Figure 12.59. Change the directions of the extrusion and/or thickness if necessary so that the extrusion is to the left and the dimensions apply to the outside of the part, as shown in Figure 12.60.

FIGURE 12.59

There are a number of tools that are specific to sheet metal parts. We can access these tools through the Sheet Metal toolbar or by adding the Sheet Metal Tools to the CommandManager. For new users, the Command-Manager helps to distinguish the tools from one another, as the tool name is displayed along with the icon.

FIGURE 12.60

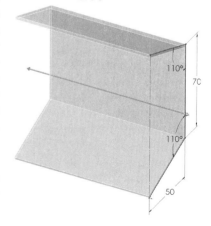

Right-click on the CommandManager and choose Customize

CommandManager. In the list of tool groups that is displayed, click on Sheet Metal, as shown in Figure 12.61. Click away from the menu to close it.

The Sheet Metal Tools are now included in the CommandManager.

Select the Sheet Metal icon from near the left side of the CommandManager. Select the Insert Bends Tool.

Select the middle face as the one that will remain flat during the bending, as shown in Figure 12.62. Set the radius to 4 mm, and set the K-factor to 0.50. Leave the auto relief checked with the type as rectangular and the factor set to 0.50, as shown in Figure 12.63.

The resulting bent geometry is shown in Figure 12.64.

The K-factor is used in converting the bent geometry into the flat geometry. A K-factor of 0.5 means that the length of the flattened metal will be calculated based on the arc length at the midthickness of the metal. This is typical for relatively large bend radii. If extremely tight bends are desired, then the K-factor may need to be adjusted to get a better correlation between the flat geometry and bent geometry dimensions.

The FeatureManager now shows several new items, as shown in Figure 12.65. These new items include Sheet Metal, where the bend radii, K-factor, and Auto Relief factors are defined, Flatten Bends, where individual bends can be edited, and Process Bends, which restores the bends when the part is rebuilt.

The part can now be toggled between the flat and bent configurations using the Flatten Tool.

FIGURE 12.61

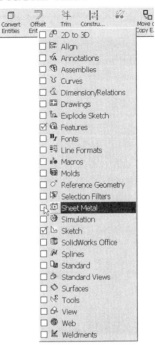

Sheet Metal

Insert Bends

FIGURE 12.62

FIGURE 12.63

FIGURE 12.64

FIGURE 12.65

Click on the Flatten Tool to show the part in the flattened state, as shown in Figure 12.66.

Notice that the rollback bar automatically moves when the part is flattened and/or the bends are suppressed, as shown in Figure 12.67.

FIGURE 12.66

FIGURE 12.67

Click the Flatten Tool again to display the part in the bent state.

With the part in the bent configuration, the bends themselves can be suppressed with the No Bends Tool. This tool allows you to toggle between the state with sharp corners and the state with the bends shown. Again, the rollback bar automatically moves when the No Bends Tool is selected.

To add the tabs, we need to create a new plane. Since the tabs are to allow the part to mount flush to another surface, the new plane needs to correspond to the front edge of the part.

Select Insert: Reference Geometry: Plane from the main menu or the Features group of the CommandManager. If there are any items already selected, clear them by right-clicking in any white space and selecting Clear Selections.

Select the edge shown in Figure 12.68, and then one of the front corner points, as shown in Figure 12.69. Click the check mark to create the plane.

View the plane from the right view to make sure that it is in the correct location, as shown in Figure 12.70.

FIGURE 12.70

FIGURE 12.68   **FIGURE 12.69**

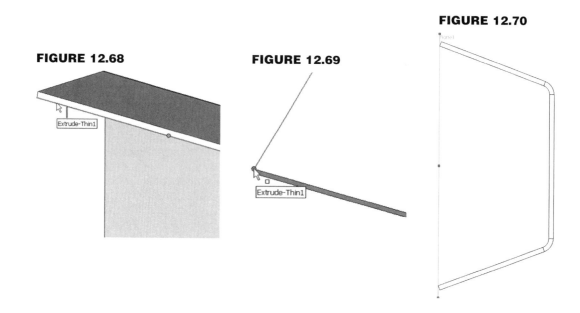

Select the new plane. Sketch and dimension the first tab, near the upper-left corner of the part, as shown in Figure 12.71.

Add centerlines, snapping to midpoints on the edges. Use the Mirror Tool to place the other three tabs, as shown in Figure 12.72.

FIGURE 12.71

FIGURE 12.72

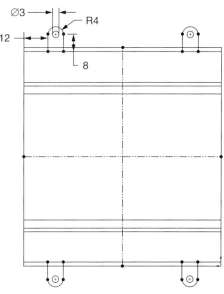

Extrude the tabs toward the rear of the part. Check the "Link to thickness" box, shown in Figure 12.73, to set the thickness as the same value that was chosen when the part was defined as a Sheet Metal part.

FIGURE 12.73

Note that the tabs have sharp corners rather than bends, as shown in Figure 12.74. This is because the tabs were added after the bends were added in the FeatureManager. Rather than define a new set of bends for the tabs, we can rearrange them in the FeatureManager. Plane1 must also be moved, since it must be defined before the tabs can be added.

FIGURE 12.74

Click with the left mouse button and hold on the new plane (Plane1) in the FeatureManager. Drag the cursor up until the arrow points between the Extrude-Thin1 and the Sheet-Metal1 definitions, as shown in Figure 12.75.

FIGURE 12.75

- Origin
- ⊕ Extrude-Thin1
- Sheet-Metal1
- ⊕ Flatten-Bends1
- ⊕ Process-Bends1
- ◇ Plane1
- ⊕ Extrude1

Release the mouse and Plane1 will now appear before Sheet Metal in the FeatureManager, as shown in Figure 12.76.

FIGURE 12.76

- Origin
- ⊕ Extrude-Thin1
- ◇ Plane1
- Sheet-Metal1
- ⊕ Flatten-Bends1
- ⊕ Process-Bends1
- ⊕ Extrude1

Now click and drag the tabs (Extrude1) to immediately after Plane1, as shown in Figure 12.77.

The tabs now appear with bends, as shown in Figure 12.78. The metal around the bends was cut per the Auto-Relief parameters set earlier. (The width of the cuts is 0.5 times the thickness, and the cuts extend 0.5 times the thickness beyond the end of the bends.)

The cut-out in the part must be added in the flattened state, so that it can be cut as a circle.

Show the part in the flattened state, and on the front surface sketch and dimension the 25-mm circle as shown in Figure 12.79, using centerlines to place the circle.

FIGURE 12.77

- Origin
- Extrude-Thin1
- Plane1
- Sheet-Metal1
- Flatten-Bends1
- Process-Bends1
- Extrude1

FIGURE 12.78

FIGURE 12.79

⌀25

FIGURE 12.80

FIGURE 12.81

- Extrude-Thin1
- Plane1
- Extrude1
- Sheet-Metal1
- Flatten-Bends1
- Cut-Extrude1 ←
- Process-Bends1

Cut the hole with a through-all extruded cut. Show the part in the bent shape by toggling the Flatten Tool. The finished part appears in Figure 12.80.

Note that the cut-out (Cut-Extrude1) appears in the FeatureManager between Flatten-Bends1 and Process-Bends1, as shown in Figure 12.81, since the cut-out was defined in the flat state.

Problems

P12.1   Open the interim assembly of the mold base and card holder that was created in section 12.2 (Figure P12.1). Perform an interference detection on the assembly. How are the interferences found related to the shrinkage factor used to create the cavity?

FIGURE P12.1

P12.2   Create a two-piece mold for the flange part of Chapter 1, allowing a 2% shrink factor. Figure P12.2a shows the lower mold half. Note that the material forming the holes is contained in the lower mold half. Figure P12.2b shows the upper mold half. Note that the chamfer feature is included in the upper mold half.

FIGURE P12.2a   **FIGURE P12.2b**

Hint: The easiest way to cut away material from the mold halves is to use *revolved cuts*. Figure P12.2c shows the sketch defining the "cutting tool" to be revolved around the centerline to create the lower mold half.

FIGURE P12.2c

P12.3 Create an assembly of the two mold halves of P12.2 and the flange. Show a section view of the assembly, with the mold halves transparent, as shown in Figure P12.3.

FIGURE P12.3

P12.4 Create a model of the sheet metal part shown in Figure P12.4. The metal thickness is 0.060 inches, and the bend radii are 0.125 inches. All dimensions shown are in inches.

FIGURE P12.4

P12.5   Create a model of the sheet metal part shown in Figure P12.5. The metal
thickness is 0.060 inches, and the bend radii are 0.125 inches. All dimensions
shown are in inches.

FIGURE P12.5

P12.6   Create the box shown from the parts created in exercises P12.4 and P12.5. Since
the sides of the box are mirror images of each other, it will be necessary to have
two different part files. Open the part created in P12.5, and select one of the
vertical faces. Then select Insert: Mirror Part from the main menu. Click the
check mark, and a new part file is created, linked to the original file. Save this
file with a new name, and then assemble the three parts to form the box.

FIGURE P12.6

The Use of SolidWorks to Accelerate the Product Development Cycle

Introduction

Any company that designs and/or manufacturing products has seen an increase in global competition over the past decades. To remain competitive, there is great pressure to develop new products. As a result, product models have shorter lives, and development costs are spread over a fewer number of units.

With these increased pressures, there has been a greater emphasis on improving and accelerating the development process. While different industries and companies have their own unique procedures, there are some activities that are typically a part of a modern product development cycle:

- Physical prototypes are used early and often.
- Computer analysis is used extensively to complement physical testing.
- Engineering functions are performed simultaneously as much as possible, necessitating better teamwork and data sharing.

Solid modeling is an important tool in the product development process. The solid model becomes the common database used for a variety of engineering functions. In this chapter, we will explore two of the most common uses of solid modeling in product design: rapid prototyping and finite element analysis. We will also consider the challenge of managing and controlling the data produced during the product development cycle.

In this chapter, you will:

- be introduced to the most popular rapid prototyping processes,
- see how the stereolithography file is used to define a solid part,
- learn how to create a stereolithography file,
- be introduced to the capabilities and limitations of finite element analysis, and
- learn about Product Data Management software.

13.1 Rapid Prototyping

The introduction of solid modeling made possible a new industry called Rapid Prototyping (RP) or Solid Freeform Fabrication (SFF). Rapid prototyping refers to the creation of physical models directly from an electronic part file using an *additive* process.

That is, the model is created by building it layer by layer rather than starting with a block and removing material to create the finished shape.

The first commercial RP method to be introduced was stereolithography in 1987. In this process, a *photopolymer,* a liquid resin that cures under the application of certain wavelengths of light, is cured layer by layer by a precisely guided laser beam. The stereolithograhy process is illustrated in Figure 13.1. The part is submerged in a vat of photopolymer, supported on an elevator platform. The fabrication of each layer begins with the elevator lowering by the thickness of a layer, typically about 0.005 inches. The part is then covered by uncured resin. A mirror system directs the laser beam across the area to be cured for that particular cross-section. Laser power and the speed of the beam across the resin surface are controlled to ensure a complete cure of the resin while maximizing the speed of the process. When all of the layers are completed, which is typically several hours after the beginning of the build,

FIGURE 13.1

the elevator lifts the essentially finished part out of the resin. Cleaning the part with a solvent and curing in a UV oven to complete the cure of the photopolymer are usually the only steps necessary to finish the part. A stereolithography machine is shown in Figure 13.2.

In the early 1990s, many new RP processes were commercialized. Among the most important was Selective Laser Sintering. In this process, plastic power is *sintered,* or fused together, by the heat from a precisely guided laser beam. The advantage of this process is that many engineering plastics, including nylon and polycarbonate, can be used. Other processes that achieved commercial success included Laminated Object Manufacturing, a process that creates parts from layers of paper, and Fused Deposition Modeling, a process that can be thought of as precisely directing a fine hot-glue gun to build up a part. The introduction of Fused Deposition Modeling was

FIGURE 13.2

Paul Roberts; Milwaukee School of Engineering

especially significant in that no laser or hazardous materials were required, allowing the machine to be placed in an office environment.

When RP was introduced, the available modeling materials had low strength and were extremely brittle. As a result, the only practical use of RP was to produce visual aids.

This use was important, since the effective communication of complex designs can minimize errors in the interpretation. However, improvements in materials have allowed RP models to be used for many other functions. The Wohlers Report[1], an annual survey of the RP industry, found the uses of RP models in 2002 to be widely varied, and included:

Functional models	22.4%
Visual aids for engineering	15.3
Fit/assembly models	15.0
Patterns for prototype tooling	10.6
Patterns for cast metal	9.2
Tooling components	6.2
Visual aids for toolmakers	4.9
Models for proposals	4.0
Direct manufacturing	3.9
Other	8.5

The fact that an RP model can now be used in a functional prototype product or in tooling (which usually requires the longest lead time of all development activities) has increased the usefulness of RP greatly over the past few years.

One of the biggest obstacles to increased use of RP has been cost. The systems described above range in cost from about $100,000 to well over $500,000. When the costs of maintenance, materials, and a trained operator are added, the costs of producing even a small RP model can be in the hundreds of dollars. Also, even *rapid* prototyping systems can be too slow at times. An engineer who has an idea for a new design concept in the afternoon and a design meeting the next morning probably will not be able to have a physical model in time for the meeting.

A new generation of RP machines has focused on lower costs and higher speeds, with some sacrifices made to accuracy and durability. The combination of low-cost materials and inexpensive off-the-shelf printing technology results in a fast and affordable (these machines costs about $30,000 or less) prototyping process.

Most domestic RP machines accept as input a type of file called a stereolithography (.stl) file. The structure of an .stl file is quite simple: the surfaces of a solid model are broken into a series of triangles, the simplest planar area. Each triangle is defined by four parameters: the coordinates of each of the three corners, and a *normal vector* that points away from the part, identifying which of the two sides of the triangle represents the outer surface of the part.

To illustrate how an .stl file defines a solid part, consider the simple part shown in Figure 13.3. Since this part has only flat, rectangular surfaces, two triangles can exactly define each surface, as shown in Figure 13.4. Since the part has six surfaces, the part can be described by 12 triangles.

FIGURE 13.3

FIGURE 13.4

[1]*Wohlers Report 2003, Rapid Prototyping, Tooling & Manufacturing State of the Industry Annual Worldwide Progress Report* (Fort Collins, CO: Terry T. Wohlers, Wohlers Associates, Inc.; wohlersassociates.com).

FIGURE 13.5

To illustrate how an .stl file is created for a part with curved surfaces, consider the flange part modeled in Chapter 1, which is shown in Figure 13.5.

To create an .stl file from a SolidWorks part, select File: Save As from the main menu and select STL File as the type of file from the pull-down menu, as shown in Figure 13.6.

Before saving the file, select the Options button. This opens the Export dialog box, and allows you to adjust the resolution of the .stl file, as shown in Figure 13.7. If the Preview option is selected, then the triangles of the .stl file are shown displayed in the part window.

FIGURE 13.6

The triangles created for a resolution set to Coarse are shown in Figure 13.8. Note the rough approximation of the fillets by the triangles.

FIGURE 13.7

FIGURE 13.8

The number of triangles created and the file size are displayed in the Export dialog box. For the Coarse resolution, 2120 triangles are used, and the size of the .stl file is 106 kb.

If the resolution is changed to Fine, then many more triangles are added to the curved surfaces, resulting in a smoother approximation of the actual shape, as shown in Figure 13.9. In this example, 5648 triangles are created, resulting in a file size of 282 kb. This file size is small enough to be easily transferred by e-mail or floppy disk.

Usually, the Fine quality is sufficient. For parts where greater resolution is desired, the Custom quality can be selected, and the slider bars moved toward the right. The only trade-off for this higher

FIGURE 13.9

FIGURE 13.10

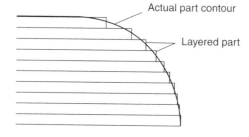

Actual part contour

Layered part

quality is larger file sizes. However, since the transfer of large files via e-mail or Internet has become easier in recent years, file size is usually not a major problem.

When the .stl file is received by the RP machine, the file is "sliced" into cross-sections by machine-specific software. If the part contour is curved, then the surface finish will include "stair-stepping," as illustrated in Figure 13.10. Some RP machines allow the layer thickness to be adjusted, but, of course, thinner layers result in increased build times.

A model of the flange created by stereolithography is shown in Figure 13.11.

FIGURE 13.11

Paul Roberts; Milwaukee School of Engineering

13.2 Finite Element Analysis

Finite Element Analysis (FEA) is a method of predicting the response of a structure to loads by breaking the structure down into small pieces (*elements*). The locations where the elements come together are called *nodes*. Equations predicting the response of each element are then assembled into a series of simultaneous equations. Solution of this system of equations yields the displacements of the nodes. From these displacements, the *stresses* (forces per unit area) are calculated for each element. From the stresses, the factor of safety against failure of the structure can be predicted. The finite element method can also be applied to fluid flow analysis, heat transfer, and many other applications.

Solid modeling has enabled the increased usage of FEA, because it automates the task that in the past required the most time: creating the geometric model and the finite element mesh. From the solid model, a few mouse clicks are all that is required to produce the finite element mesh. Figure 13.12 shows a mesh created from the bracket model of Chapter 3. A 100-pound load will be applied to the end of the flange. After the mesh is created, the loads and *boundary conditions* are applied. Boundary conditions are the displacements controlled by external influences. For example, if we assume that the bolts attaching the bracket to the wall cause the back face of the bracket to be perfectly fixed, then we would apply corresponding constraints to the movements of the nodes on that face. The stresses resulting from the 100-pound load are shown as stress contours in Figure 13.13.

FIGURE 13.12

100 lb

As FEA has become easier to use, it has also become easier to *misuse*. While the software takes care of the number-crunching, a knowledgeable engineer is required to set up analysis and interpret the results.

Among the common errors made by inexperienced users of FEA are:

- *Inappropriate choice of elements.* Solid modeling software with built-in FEA meshers allows a mesh of solid elements to be easily created from a solid model. However, in many cases

FIGURE 13.13

Stress
von Mises
lbf/(in^2)

8000
7000
6000
5000
4000
3000
2000
1000
0

solid elements are not the best choice. Relatively thin structures subjected to bending loads are usually better approximated with shell elements. A frame made up of welded structural members may require tens of thousands of solid elements to produce an acceptable grid, while a few beam elements will produce better results with substantially less calculation time.

- *Use of only linear behavior.* Linear analysis is based on the assumption that structural response varies linearly with loading. For example, the deflections and stresses of the bracket when loaded to 200 pounds are exactly double those produced by a 100-pound load. There are two types of nonlinearities possible, however. Geometric nonlinearity is present when the structure's stiffness is significantly different in the displaced configuration than it is in the original configuration. Material nonlinearity is present when the material is stressed beyond its *yield point.* When the material has yielded, it may not break, but will deflect farther with little or no additional loading. Both of these nonlinearities require *iterative* solutions. That is, rather than applying all of the load at once, an increment of the load is applied and the equations are solved. The equations are then reformulated based on the results of the first solution, and solved again. This process is repeated until the solution is found. While this procedure requires much more processing time, it is necessary to obtain an accurate solution for many problems.

- *Inappropriate boundary conditions.* In the example of the analysis of the bracket, it was assumed that the back face of the bracket was fixed to a rigid wall. Is this a reasonable assumption? Maybe, but if the screws holding the bracket stretch slightly, then the top of the bracket will separate from the wall, placing more force on the bottom of the bracket. Most actual structural restraints are neither perfectly fixed nor perfectly free, requiring the engineer to use judgment in specifying the conditions to be placed on the model. In some cases, the analysis may be performed several times with different boundary conditions to obtain the limits of possible structural response.

- *Misinterpretation of results.* In Figure 13.13, the stresses displayed are the von Mises stresses. These stresses are calculated based on a specific failure criterion (the von Mises criterion) that is an excellent predictor of the yielding of ductile materials, but it may be inappropriate for predicting the failure of many materials.

These common errors are not presented to discourage the use of FEA, but rather to encourage the proper use of the method. An engineer should have a good understanding of mechanics of materials and at least an introduction to FEA theory before using FEA for any important application.

13.3 Product Data Management

We have mentioned earlier that solid modeling supports concurrent engineering, since many engineering functions can work from the same database (the solid model). Allowing multiple users to work on the same model or drawing, however, creates challenges, as well. How do companies manage their part files and drawings to control who can make changes?

Before answering this question, it is helpful to consider how companies managed paper drawings and other data before solid modeling was introduced. When a draftsman completed a drawing, it was checked and then went through a *release* process in which it was approved and signed off by various department representatives (design engineering, manufacturing engineering, safety, etc.). Copies were made and distributed, and then the released drawing was stored in a vault. Most medium-sized and large companies had a *configuration management* department that controlled the release process and subsequent changes. When a change was requested, a formal engineering change order process was followed. When the change was approved, it was amended to the drawing, or the drawing was modified and given a revision number, if the change was significant. It should be noted that many drawings would eventually have dozens of associated change orders, so managing this process was an important job. When a user of the drawing, say a stress analyst about to begin an FEA model, needed a copy, the configuration management department provided the latest version, including all change orders. Similarly, the Bill of Materials for each part was maintained and provided to the purchasing and manufacturing departments. In addition to controlling the drawing changes, the configuration management department would maintain *drawing trees* that showed how drawings related to each other. A change in a part could affect an assembly at the next level. Other data might have been controlled but not necessarily linked to the drawings. For example, a stress analysis report might have been released and stored in the vault, but a person looking at the drawing would probably not know that the analysis had been done, or where to find it. Similarly, in industries where weight is important, mass property reports were produced, but usually not linked to drawings.

The introduction of solid modeling presented many new challenges to configuration management. Some of the features that make solid modeling so exciting to design engineers—associativity between parts, assemblies, and drawings, ability to use the solid model for many functions—could cause huge problems from a data management perspective. That is, one person could make a seemingly simple change to a component and that change would propagate throughout an entire assembly without the person being aware of its effect. Now that some companies model major systems with solid modeling (Boeing modeled the entire 777 aircraft), controlling who can make changes is vitally important.

Product Data Management (PDM) software is used to perform some of the tasks of the configuration management department, while streamlining communications between various departments. PDM entails two broad categories of functions:

- Data management, the control of documents (part files, drawings, stress reports, etc.), and
- Process management, the control of the way in which people create and modify the documents.

The data management function is similar to that of the paper-based configuration management department, except that the released drawings are now electronic files and are stored in a *virtual* vault instead of a physical vault. (Actually, backup tapes and disks of the virtual vault are often stored in a fire-proof physical vault.) Since the part and drawing information is stored in a relational database, immediate location and retrieval of files is possible. This helps to reduce redundancy, especially among standard components such as rings and fasteners. A design engineer who needs to specify a fastener can easily determine if there is already a similar fastener with a part number assigned in the system. If there is, then the purchasing department will not have an additional item to buy.

Process management is the control of active procedures: who generates the data and how the data are transferred from one group to another. One important feature of PDM systems is work history management. In a paper-based system, having old, outdated drawings around is an invitation for trouble, since they can be used by mistake. But by destroying old drawings, the history of the modifications made to that drawing can be easily lost as well. PDM software can track the change history of a part, an important function in Total Quality Management (TQM), while protecting against accidental usage.

What does the future hold? Most manufacturing companies implemented Manufacturing Resource Planning (MRP) systems long before PDM became popular. MRP systems allow the tracking of raw materials, work in progress, and finished inventory in the plant. The move to reduce inventories and adopt just-in-time raw materials deliveries necessitated the adoption of MRP systems. In many ways, implementing PDM to engineering functions is analogous to implementing MRP for manufacturing.

The logical next step, then, is to combine MRP and PDM systems into a single system referred to as Product Lifecycle Management (PLM) or Enterprise Resource Planning (ERP). As with any new large-scale system, implementation is expensive and time-consuming, and so these systems are not yet in widespread use.

13.4 Some Final Thoughts

The widespread adoption of solid modeling has been part of a revolution in the way that products are designed and developed. As many companies have thrived with new technology, others that have not kept up have not been able to compete and have been forced to close. Although no one can predict the future, one thing seems certain: the pressure on companies to develop new products faster and better will not lessen.

On a more personal scale, the same concept has held true for many engineers. An engineer who is adverse to change and reluctant to learn new tools is at a great competitive disadvantage to his or her peers.

The good news is that most people who enter the engineering profession do so because they have the curiosity to want to learn new things. Keeping that curiosity alive will allow an engineer to have a rewarding career that is always fresh and interesting.

Index